A Hand Book of

Seed Industry

(Prospects & its Costing)

Avinash Ramdeo

SCIENTIFIC
PUBLISHERS (INDIA)
www.scientificpub.com

Reprint : 2021

Publihsed by:
Scientific Publishers (India)
5-A, New Pali Road, P.O. Box 91
Jodhpur - 342 001 (Raj.)
E-mail: info@scientificpub.com
www.scientificpub.com

ISBN: 978-81-7233-677-6

Lasert typeset : Rajesh Ojha
Printed in Inida

This book has been
dedicated to
my Dada Guru Shri Ramsingh ji Guru
God Shri Chandra Gupta,
Shri K.K. Gupta,
Grand Father Shri Inderraj ji Ramdeo and
Grand Mother Smt. Laxmi Devi.
By blessings of
my Father Shri S.S. Ramdeo,
Mother Smt Shanti and Shri P.K. Gupta
I have tried my best to compile this book

Preface

After joining Rajasthan State Seeds Corporation Ltd. at Seed processing Plant Durgapura Jaipur in 1985 under the Plant Incharge Sh Chanchal Chand Khivsara. The work for processing of seed was allotted to me.

In 1989 I was posted at Head office Rajasthan State Seeds Corporation under Dr. S.C. Mathur Dy. Manager Processing. Those days the report of Mathuradas committee was in the news and in which all the corporations, which were in losses, were recommended for closer, the name of my corporation was also there. This pinched me very much that if corporation will close what will I do then? In the Head Office I drew the attention toward the actual costing incurred in the processing and packing. Mrs. Shashi Mathur Financial Advisor and Smt. Reeta Sharma Astt. Manager Finance encouraged me to work on the above matter.

Although I tried my best to work upon to the best of my knowledge, but all the aspects might not have been covered. Though you can get the overall view of the Seed Industry.

Former Managing Directors Shri R.R Metha Shri C.S. Rajan Mrs. Shashi Mathur RAcS (Former Sr Manager Finance RSSCL), Mrs. Reeta Sharma (Former Astt. Manager Finance RSSCL) by the grace of the above official's Blessings I was able to come up with this matter in your hands.

I am heartly thankful to Mrs Malavika Dadlani, Dr Purshottam Chandra IARI New Delhi, Shri N.S. Rathore, Shri B.D. Bhargava (Ex Director RSSCA Jaipur), Shri Mustansir, Regional Manager NSC Jaipur, Dr. P.K. Dadheech. Jt. Director (Botany), Seed Testing Lab Jaipur, Dr. Mahesh Shrimali, Asso. Prof, Dr. Shaitan Singh Shekawat, Dr. S.S. Bareth, Astt. Proffessor ARS Durgapura, Shri Abdul Khan AARO Seed Testing Lab Durgapura, Jaipur.

I am also thankful all the Directors Seed Certification Agencies who sent me the addresses and codes of registered Seed Processing

Plants on my request, all my colleagues and officers who guided me from time to time.

The interpretations drawn are based on my small knowledge. One cannot be the expert though some error might be there as this is my first experience I hope you will ignore but valuable suggestions are always invited.

Avinash Ramdeo
B-32, Ganesh Marg
Bapu Nagar,
Jaipur 302 015
E-mail: avinashramdev@gmail.com

Contents

INTRODUCTION

N.S. Rathore & Avinash Ramdeo

Prospects of Seed Industries

Seed is the most important and primary Agriculture input. Its quality has a direct impact on production. The need of quality seed has been felt since times immemorial when our farmers thought of selecting good fruits, seeds and planting material to raise healthy crops. The Royal Commission on Agriculture 1926 and the National Commission after Independence have strongly emphasized for need of quality seed to feed the fast growing population.

India in an Agriculture based country. It is a well known fact that 68% of the population is still engaged in agriculture. Unfortunately the ratio between the growth rate of population and the increase in production of food is very low. The situation is quite alarming and will become more critical after 20 years. If we do not target the higher productivity resulting in a many fold increase in overall production. The Nation has to keep the close watch on the use of quality seed. It certainly needs a joint effort at level of scientists to develop location specific high yielding varieties providing better crop geometry and other agronomic practices where the farmer should be educated to utilize the maximum of new technologies.

The Government sector also has to be vigilant enough to ensure quality inputs to farmers and ensure better marketing services A well developed seed Industry as such can be the only answer to ensure higher production and better economy of Indian farming.

The development of seed multiplication Farms during late 50's, and the thought to establish National Seeds Corporation, State Farm's corporations etc during 60's were the major steps to strengthen the seed Industry in the country. The National Seed Project and the encouragement to private sector seed trade are the further steps taken by the Government in this direction; therefore today we have the well-developed Seed Industry in existence. We are fully prepared to enjoy the fruits of Globalization where High-tech Agriculture is just preparing to step into our farmer's fields. Only what is needed is to keep our eyes open with a receptive mind in such a way that the poor farmer is attached only to real Diamonds, Imitations are identified and discarded out rightly before they reach to the poor farmers.

At the time of Independence food grain were approximately 51 million tones feeding a population of 36 million people comparing it to years 1983-84. India produced over 150 million tones of Grains, with feeding population of approximately 100 million people. An Australia is borne in India every year. The RATIO OF INCREASING POPULATION TO THE NATION'S FOOD GROWTH PRODUCTION IS TOO LOW. Which is a serious question for Indian economy.

"We look the Brighter Future for Seed Industry in India."

Chapter-1

Some important milestone of development of Seed Industry in India

1937 Imperial Council of Agriculture Research was formed (ICAR). Main target was to increase Jute and Cotton Seed Production, as it was main crop those days so to gain money.

1945 Royal Commission on Agriculture formed.

1952-55 Plan for state level seed production through registered seed growers.

1950 ICAR started to generate new variety of prevailing crops and started State level research through Research Stations with the coordination.

1955-57 ICAR brought the concept of 1 Seed of Crop; 2 Seed of variety.

1957 All India maize Coordinated Project was started under the supervision of ALL India Crop Improvement project on Maize PL-480 Scheme from USA. Then Ganga-1 and Ranjit were taken for further development.

1957-61 It was the turning point in Seed development in India. Under the guideline of A.A. Johnson Agriculture production Team laid group seed Production plan.

1.7.1963 National Seeds Corporation was established with the view of producing and marketing of available foundation seed to extend the certified seed promotion activities of seed production with registered Seed Producers.

1965-66 This was also the next turning point of seed development in India. The high Yielding variety programme (HYVP) was launched by Shri Subramanyam, Shri M.S. Swaminathan and Shri Shivraman.

1965 18000 qtls. Mexican Seed was imported to India to increase seed production. The seed was free from all type of bacteria.

1966 Seed Act came into action.

1968 Seed review team gave the seed review report.

1969 Central Seeds Certification Agency was established to assist the seed production in India with the Joint Working Party (JWP) & State level Seed Corporation were started. Simultaneously Agriculture University, Seed Testing Laboratory were established.

As it is said earlier that the basis of seed industry is seed, and looking towards the concept "better the seed best will be the produce". Looking to the population growth to production ratio the world wide programme were made to cover up the requirement of food grain through World Bank guidelines in India.

Firstly the Seed Industry was started in the states as State Seeds Corporation under National Seeds Corporation upgraded though National Seed Project in 1976-80.

Under NSP-I the following seed corporations were established in following States under the guideline of National Seeds Corporation Ltd.

State	Type of Investment
Punjab	Farm Development
	Seed processing Plant
	Cotton Seed Processing Plant
Haryana	Farm Development
	Seed processing Plant
	Cereal Seed Processing Plant
Maharashtra	Farm Development, Seed Processing Plant, Cereal Seed Plant, Cotton Seed Processing Plant, M.S.S.C.
Andhra Pradesh	Farm Development, Seed Processing Plant, APSSDC, Cereal Seed Plant
Rajasthan	Farm Development, Seed Processing Plant, RSSC (1) SGNR (2) Suratgarh (3) Kota SFCI.

Uttar Pradesh	Farm Development, Seed Processing Plant
Bihar	BSSDC, Seed Processing Plant.

On success of NSP-I the following project were also established under NSP-II in the following States.

Orissa	OSSC
Karnataka	Farm Development, Seed Processing Plant, KSSC.
1988	Under Shri Rajeev Gandhi then Prime Minister of India a new policy to Import of Vegetable Seed to India was enforced and Import was liberalized. The Indian Farmer would Import any vegetable seed from any place

Chapter 2

Seed and Plant

"पौधे धरती का श्रृंगार है।"

INTRODUCTION

Have you ever thought how the plant or trees had been on the earth? What and how do they share with our lives? And how they are propagated?

It's true "If the word plants life is removed from human life or earth we can't survive even for seconds. Our life saver Oxygen, food, fodder, vegetables, fruits etc. all are the by-product of the plants. The plants play an important role in our life.

Farming

1. Traditional Farming : In the traditional farming farmer use to pick the good quality seed from his own field.

Parameters

1. Sowing method Chhatai

2. Scientific Farming increase production and economy of farmers two characters are combined.

METHODS OF GROWING PLANTS

(a) Vegetative method

(b) By Seed

(a) Vegetative Method

The cutting of mother plant is sown for vegetative growth to produce new generation. It is done in those plants in which there is lack of ovary or fertilization due to adverse climate etc. *e.g.* (1) Sugarcane : Cutting is used

(b) Seed

The living embryo when sown it produces true to type plants are called seed. Embryonic lige that produces a new development after it is separated from its parent is called seed. Seed is the container of Embryo plant through hazards of time and space where turn produces seed, which can regenerate true to type plants. Seed is the biological entity capable of producing its own type Wheat, Barley, Bajra, Paddy etc.

How Seed is produced

The ovule of seed develops in seed ovary of seed vessel. The ovary is the part of the flower that contains ovule or an egg cell. These cells are fertilized by the sperm cell i.e. pollen before it can particulate the parent life.

While developing an embryo these develop special store of food which feed nutrition in its turn to newly developed embryo till it starts its own feeding at the established place. It contains fats, carbohydrates and proteins, which provides initial food to the embryo.

Type of Seed

1. Monocot eg Wheat
2. Dicot eg Gram

How Seed is formed

a - petals
b - Sepals
c - epical
k - pollen
kk - pollen tube
cc - egg cell

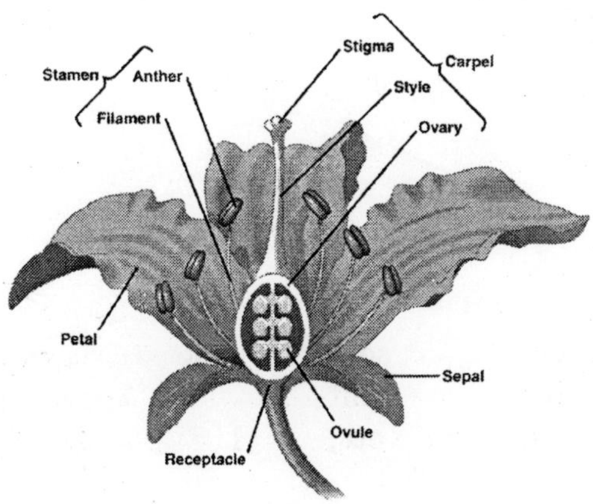

A typical type of flower, which have ovary and pollen tube. Ovary is fertilized by pollen grain and embryo is formed. We can say that the pollen is shifted to the stigma of the flower of the flower by wind, insects, and mankind form style and reached to the egg cell. It is by fertilization forms an embryo, if they are more eggs in ovary it will produce more seeds, after the fertilization of egg cell ovary develops and seed is formed.

Seed maturity and seed germination follow a sequence in the life of a plant normally seed is separated in space and time from mother plant by natural or manual.

SEED AND GRAIN

Seed: It is the container of an embryo from which a new plant can grow, flower in turn produce seed which can regenerate true to type plants it has some certain standards in regards to the minimum germination, genetic purity standards etc.

Grain : The grain are generally those seed used for feeding the human being or animals etc, they in turn cannot produces true to type generation genetically.

The seed can be used as grain but grain cannot be used as seed.

Chapter –3

Seed Industry

INTRODUCTION

Seed Industry denotes the production allotment, processing, testing, certification packing of good quality seed and finally it is marketed to the ultimate consumer i.e. farmer to produce good quality seed which in turn gives high yield.

In ancient times when farmer grew crops in the field he used to procure some part of the produce for next year. By their experience and demand of time they thought over to commercialize and scientific manner and started to pick better quality of seed from their field. This thought was further came to research and the varieties were developed and new aspect of Seed Industry was looked in the commercial manner in all over the world.

FACTORS OF SEED INDUSTRY

Seed is primary input of Agriculture. The raw produce received after harvest needs to be prepared for use as seed. Practically every seed has to be kept stored till its next session of sowing. It needs some steps to preserve its physical properties and germination potential. Preservation of its vigor and viability is almost necessity and seed man are supposed to be acquainted with the work of seed processing which includes grading,testing,treating, packing, storage, handling etc.

Processing Plant and Building

The Plant should be installed in a suitable building with proper ventilation, alternate power arrangements. The processing plant in the covered space having the facility via: - Seed platform where seed is checked in contest with visibility and vigor also it can be checked by its moisture percentage. Unit must have at least 3 godowns with separate processing space.

Godown mentioned above one must be of raw seed and another for Graded Seed and packed seed. The godown should be of finished walls free from cracks and crevices, fitted with exhaust fans, lightening arrangement and equipped with Fire extinguishers and First Aid Box.

Processing Machine and Equipments

Seed processing machine and equipments industry are the following -

(a) Pre-cleaner

(b) Grader

(c) Sieves of different size (Crop wise) [As refer in Minimum Seed Certificates - Standards]

(d) Indent Cylinder

(e) Gravity Separator

(f) Seed treater

(g) Vacuum cleaner/ Blower

(h) Weighing Machine of different size

(i) Bag Closer

(j) Moisture Meter

(k) Seed Dryer

(l) Elevator

(j) Wooden crates / or iron pallets.

After functioning the equipments these are to be kept cleaned for better result. It is possible that same seed may leave in the remote corner thus by help of Vacuum cleaner equipments are cleaned to became desirable to avoid any kind of Contamination. Oiling gracing and regular check up of bearing may help in smooth working of machine and maintains life of machine.

(a) Pre-cleaner

Pre-cleaner is the machine, which cleans the seed prier to the grading. It removes the husk gravels and the scalped material inert matter and the seed is pre cleaned. This is installed with the heavy machines whose capacity is too high.

Fig. 3.1. Pre-cleaner

(b) Grader

Grader is well-equipped complete machine which grades the raw material feed to it. There are the following parts of seed grader.

1. Feeder
2. Screens
3. Electric Motor
4. Air Separators
5. Traveling brushes

There are different type of graders i.e. 2 Screen, 4 Screen. The body of machine is of steel or of wood.

Fig. 3.2. Seed Grader

Fig. 3.3. Seed Grader

Fig. 3.4. Mobile grader

Constructional Features

1. Frame and Body : This is made-up of steel channel the whole machine is assembled in units so as to carry through narrow passage.

2. Screen : This is durable and are inter changeable according to requirement i.e. Crop wise.

3. Traveling Screen Brushes : Each screen is attached to a pair of brushes mounted on a steel carriage. These brushes move beneath each screen to remove the blockage in screen.

4. Feed hoper : It provides steady and uniform flow on the screen and is so designed that it can be flooded with seed when passing course material feed shut off clutch is provided on the feed roll.

5. Support for Screening : All spots shake with the shoes and are picked for self-cleaning.

6. Air Separation chamber : It removes all light Trash, dust, empty pods hollow kernels etc. from final Seed, and speed can be controlled as required by a knob provided on the machine body.

7. Drier : It is drier form main motor shaft either by V Belt or Roller chain.

Operational Features

The seed to be processed if feeded manually or by elevator in seed hopper from there if it falls into top screen light Impurities and dust are removed by the feed air where as heavier impurities such as stony matter etc. are removed from seed mass by force of gravity. Though feed air trunk this seed is passed over the lighter screen where weed seed immature grains are removed in two screen grader this screen is not available and but the standard seed sieve crop wise is present. The Seed has to be passed through the standard sieved. Again sound seed are cleaned by passing a flow of Air stream. These sound seeds come out by an output pipe to bag holders. A rotary blower presses the flue air containing the dusty matter into a chamber or exhaust pipes. Finally we get the uniform seed or graded seed for further testings etc.

(c) Screen/ Sieves

Crop wise Screen aperture size has been prescribed by Govt.of India in book Minimum Seed Certification Standards. There are standard sizes fixed to get the good quantity of seed crop wise. The size can be changed in consultation with the State Seeds Certification Agency to get quality seed. There are two types of screens, which are used in the seed industry.

1. Slotted
2. Round

Slotted : It means the holes in the screens are having rectangular shape.

Round: The screen holes are of Round is size.

S.N.	Name and type of seed	Aperture sizes in mm	
		Top	Bottom
	CERALS AND MILLETS		
	Bajra		
a.	HB-1, Tift of 23A, 23B, HB-3, HB-4, Tift 23 D2A, Double Dwarf Tift 23 D2B	3.25, (3.18)	1.3, 1.6 (1.27x19.05)*
b.	J- 104, K-560, BIL-3B	3.25, (3.18)	1.9, 1.4, 1.98 (1.41x19.05)*
	Maize		
a.	Hybrids		
1.	Ganga Safed-2, Hi- Starch	11.00,(11.11)	7.00 (7.14), 7.50 (7.54)
2.	Deccan, Deccan-102, G-101, Him-123, Ganga-6 Ranjeet	11.00,(11.11)	7.00, (7.14)
3.	Ganga-3	10.50,(10.72)	(6.75)
4.	Ganga-5	10.50, (10.72), 11.00 (11.11)	7.00, (7.14)
b.	Composite and open pollinated varieties		

1.	Jawahar, Vikram, Amber, Kisan	11.00, (11.11)	7.00, (7.14)
2.	Vijay	10.50, (10.72), 11.00 (11.11), (10.72)	7.00, (6.75), (7.14)
3.	Pop Corn	8.75, (8.73)	4.75, 4.25, (6.75)
4.	Sona, Puerto Rice Grade-1	10.50, (10.72)	(6.75)
c.	*Inbreeds*		
1.	CM 109, 110, 111, 202, 205, 206, 600, (open pollinated)	10.50, (10.72)	6.40, (6.35)
2.	CM 103, 104, 105, 112, 113, 114, 115, 201, 300, 400, 500, Antigua Gr-1; CM 601 open pollinated	10.50, (10.72)	(6.75)
d.	*Single Crosses*		
1.	CM 109 x 110, 113x112, 202x201, 202x205	10.50, (10.72)	6.40, (6.35)
2.	CM 104 x105, 202 x 111, 400 x 300, 115x114, 202 x 206, 103 x 104, 210 x106	10.50, (10.72)	(6.75)
	Paddy		
a.	IR-8, jaya, Pusa,2-21, IR-20, Ratna, Improved Sona, IET 1990, PR –4-14	2.8, (2.78x12.7)* (2.78x19.05)	1.85, (1.81x19.05)* (1.19x19.05)*
b.	Fine and Superfine varieties	2.8, (2.78x12.7)* (2.78x19.05)	
	Sorghum		
	CSH-1, CSH-5, CSH-6, CSH-4 & MSCK 604, CK 60 B, MS 2219 A, 2219 B, IS 3691, MS 2077 A, 2077 B, CS 3541	4.75, (4.76)	3.5, (3.57), 2.1 (2.12x12.7)* (2.12x19.05)*
	Wheat		
a.	Sonalika (HD-1553)	5.50, (5.56)	2.1, (1.95x19.05)* (2.12x19.05)*
b.	HD-1982 (Janak), HD-1981 (Pratap), HD-2009 (Arjun) and other varieties of similar seed Sizes	5.50, (5.56)	1.8, 2.0, (1.95x19.05)*
c.	Kalyan Sona, Malvika, Raj-911	5.50, (5.56)	2.0, (1.95x19.5)* (1.95 x 12.7)*
d.	Sharbati Sonara; Safed Larma, Chhoti Larma	5.50, (5.56)	2.0, 2.1, (1.95x12.7)* (2.12x12.7)* (1.95x19.05)*
	PULSES		
	Bengal Gram		
	130, C-214, 4110, H-208		5.5, (5.56)
	Black Gram (Urd)		
	B-1, Type 9, PS – 16		(2.78x19.05)*

	Crop / Variety		
	Moong / Green Gram		
	Pusa Baisakhi, S-8, PS-7, PS-10, PS-16	(5.16)	(3.18x19.05)* (2.78x19.05)*
	OIL SEEDS		
	Castor		
	NPH-1		(5.95, (4.37x12.07)*)
	Soyabean		
	Bragg, Clark-63, Lee, Hill, Hardee, Improved Pelican, Punjab No. 1	8.0, (7.94), 8.33	4.0, (3.97x19.05)*
	Sunflower		
	EC-68414, EC-58415, EC-69874	(10.32)	(4.76)
	FIBERS		
	Cotton		
a.	Acid Delinted	(7.14)	(3.97x19.05)
b.	Fuzzy	(14.29), (4.76)	7.00, (5.16x19.05)* (2.12x19.05)*
	Hemp / Mesta		
	Jute		
a.	Capsularis	2.4, (2.38)	1.60, (1.19)
b.	Olitorius	2.0, 2.4, (1.98), (2.38)	(0.79)
	FODDERS		
	Berseem		
a.	Diploid	1.9, 2.4, (1.98), (2.38)	(26x26)**
b.	Tetraploid	2.4, (2.38)	1.2, (1.19x12.7)*
	Cowpea		
	Fodder		
	EC-4216, Type-2	7.0, (7.14)	3.5, 1.2, (3.57x19.05)*
	M.P. Chari, Pusa Chari		2.1
	Oat		
	Kent &Algerian	(8.73)	(1.81x19.05)* (1.81x12.7)*
	Lucerne		
	Type – 9	(2.38)	(26.26)**
	VEGETABLES		
	Pusa Barsati	(5.95)	(3.97)
	Pusa Phalguni	(5.95)	(3.18x19.05)*
	Peas T – 163 Pulse Type	10.5, (10.72)	(6.75)
	French Beam	11.00, (11.11)	4.75, (4.76x19.05)*
	Okra (Bhindi)	6.00, (5.95)	4.35, (4.37)

Cluster Beam (Guar)		
PNB-(Veg. FS-277), (Fodder), Pusa Sona	(5.95)	(1.81x19.05)*
Pusa Sona		(1.81x12.7)*
Bottle Gourd	-	-
PSPL & PSPR, Bitter Gourd	11.00, (11.11)	(6.35), 6.5
Tomato, Chllies, Brinjal	4.0, (3.97)	2.0, (2.12)
Onion PR	(3.97)	(2.12)
Cauliflower	3.25, (3.18)	2.4*, (1.4x19.05)*
Methi Kasuri	(2.12), 2.1	(26x26)**
Methi Early Bunching	3.25, (3.18)	(1.19x7.94)*
Spinach	5.5, (5.56)	1.85, (1.81x19.05)* (1.81x12.70)*
Dolichos Bean	(8.73)	4.75, (4.76x19.05)*
Carrot		(26x26)**, (26x28)**
Watermelon	6.0, (6.35)	1.85, (1.81x19.05)* (1.81x12.70)*
Turmip	(1.81)	(1.27)
Radish-Japanese White	(3.97)	1.85, (1.81x19.05)* (1.81x12.70)*
Radish – IRRWT & PC	(3.97)	(1.41x19.05)*
Sponge gourd	9.5, (9.53)	6.4, (6.35)
Knol-khol, cabbage	(2.78)	(1.19x7.94)* (1.21x19.05)* (1.27x19.05)*
Sugar Beet	(7.94)	(3.17)
Muskmelon, Cucumber JLG	(7.94)	(3.57)
Lettuce-Chinese Yellow and Great Lakes	(1.81)	(1.27)

Note: 1. An asterisk (*) with an aperture size denotes oblong holes, two asterisks (**) denote wire mesh sieve, and the rest, without any superscript denotes round hole type sieves.

2. Figures in paranthesis are for the inch series, and given in equivalent millimeters.

Screen aperture sizes listed above should be taken for reference and guidance only. Actual screen sizes to be used for processing of certified seed of certain crop should be decided in consultation with concerned state seed certification agency.

Again sieve size can be lowered with the permission of Certification agency graded seed shall not have seed of size lower than the bottom screen size used beyond 5% (by weight).

Ref: MSCS (1988) Page No. 7 Para XIX

(d) Indented Cylinder

When there are cut grain or weed seed is present in the in the crop Indented cylinder is required to separate these from seed at seed processing plant. Indented Cylinder is of generally two types -

1. One Cylinder
2. Two Cylinder

It is a steel construction with variable cylinder speed. It can also be adjusted for speedy discharge or rejections.

(a)

(b)

(c)

Fig. 3.5.

The final graded seed received from grader is put into the Indented Cylinder by means of hopper, Elevator etc and the make is such that it has round grooves in the inner roater and while rotation the cut grain and weed seed is hold-up in the grooves, this is put off on the outlet panel to come out separately from seed thus it removes the cut grains and weed seed from graded material and we get the finally pure (free from other inert or cut material) Seed.

Single Cylinder are generally attached with the grader but two Cylinders device is attached/fixed aside the machine. This machine is specially designed for removal of broken/cut seed and weeds e.g. Hierankhuri and cut seed from Wheat.

(d) Gravity Separator

It is all steel construction, prompt monitoring through centrally located control for feed rate, deck oscillation speed, side and end deck indignation and independent regulation of each fan. It separates seed by gravity so as to get uniform seed. It is available in 3 fans and 5 fans models some seeds have heavier particles like stones etc. of same size it separates them from seed by working on the principal specific gravity and fluidization. The seed passes through an oscillation deck with a continues airflow. The seed spread allover entire deck it is practically so raised from the deck floor that the lighter once move to one side and the heavier seed tends to move on other side of the deck. This equipment even separate out the insect damaged seed. Cut seed, lighter seed gravels, stones form seed. It works on the basis of the gravity of the material i.e if the màchine is adjusted on the seed weight of any crop than heavier or lighter particles than the seed weight is separated from the seed.

Fig. 3.6. Gravity Separator

(e) Seed Treater

The Seed Treater is equipment by which the seed is automatically treated. It is also known as slurry treater. The required ratio of water is mixed with the chemical by which seed is to be treated and the container of the treater is filled with the slurry. The system works on the basis of rotation of cups full of chemical by weight adjustment. The graded seed is put into the receiver, by adjustment the seed falls down into the rotor same time cup filled up of chemical (Slurry) is dropped through chain in the rotor. It makes the circle method for mixing the chemical each time by weights rotor pins rotates the seed and chemical is mixed with seed.

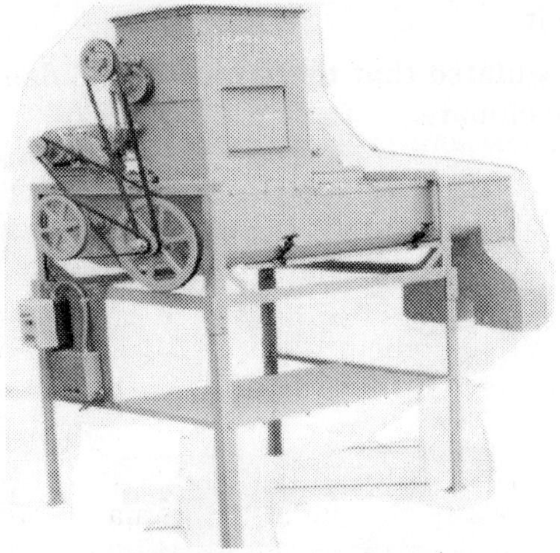

Fig. 3.7 Seed Treater

(f) Vacuum Cleaner/ Blower

It is the device by which machine is cleaned by creating air flow Vacuum and all the dust particles and seed of different varieties in remote area are sucked and machine is cleaned for next lot or variety. These are of different types it depends upon the size of machine and processing plant.

Fig. 3.8. Vacuum Cleaner/ Blower

(g) Seed Dryer

Like Hair driver the seed driver is also there to get free seed from excess moisture. It is only used when there is more humidity or continues rains etc. the humidity is constant high in Air.

There are same chambers in which the bags of seed are kept and heated air is passed through it.

This heat is so regulated that there must be no damage to seed they are mainly used in humid climates.

Fig. 3.9. Seed Dryer

(h) Weighing Machines/ Balances

Different types of weighing machines are used in Seed Industry. These weighing machines/ balance are of many types.

Machine should weigh from 5-10 grams to quintals for higher weights these balance are of following types -

1. Weigh Bridge
2. Tripod Balance
3. Platform Balance
4. Pan Balance

These balance must be regularly check up by the operator. Please note that the weights of balances must be well stamped by Weights and Measure department every year:

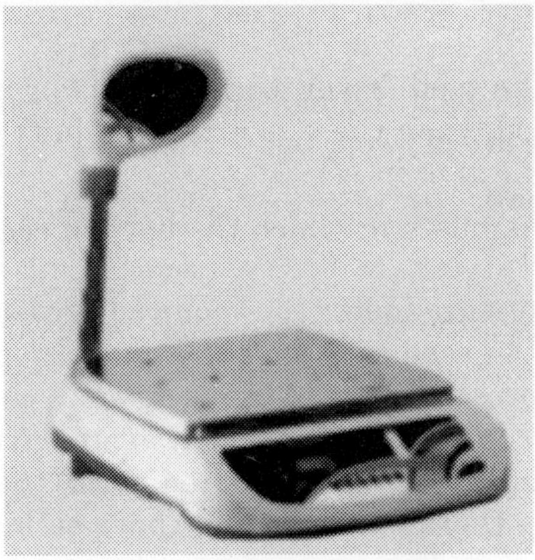

Fig. 3.10. Weigh Balance

(I) Bag Closer

The hand machine, which closes the bags or sew the seed bags is called bag closer machine.

Fig. 3.11. Heavy duty Bag closer

Thread spool is fixed on the bag closer machine and stitching is directly done by it.

These are of two types :

1. Heavy duty: It is used where continues load word.

2. Light duty: It is used where continues work is not there.

The seed bag is such sewn be chain that one cannot open bag for starting. Bags can only be opened from last closing.

(j) Moisture Meter

The equipment by which the moisture of the seed it tested is called the moisture meter. These are of two types -

(a) Electric (Manual)

(b) Digital Manual

Electric : It is moisture meter the required pressure is given to the seed in the cup i.e. as per the seed crop wise the quantity prescribed is filled into the pressure pan the prescribed pressure is given to the seed and the current is passed. First caliber is adjusted and than again current is passed through it. The reading of the meter is compared with the temperature on the moisture meter having thermometer again this reading is coincided with the chart or dial on the meter. The arrow shows the moisture content of the seed in some commodity the final reading is multiply by quiescent.

Digital : In this moisture meter the prescribed seed volume is taken and put into pressure pan the code is given as per crop. The required pressure as per chart is given. Now the Button is pressed the Digital number will appear on the board Directly and will go down to 0% and than the exact reading will appear. This will be the final moisture percentage of the commodity or seed.

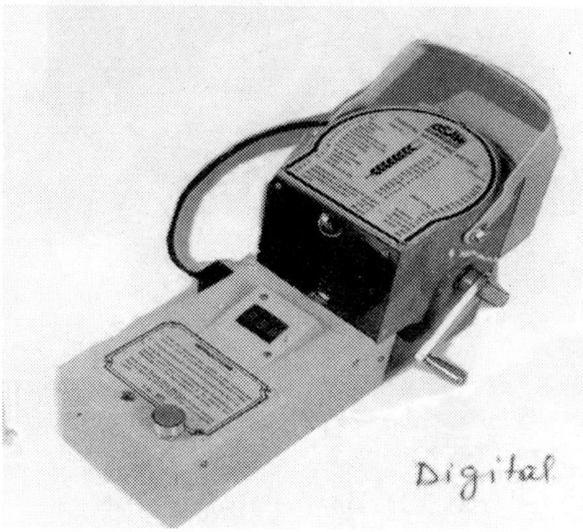

Fig. 3.12. Digital manual

(k) Elevator

These are used for physical movement of seed to different heights a vertical belt is fitted with small cups and powered to move like a Persian wheel to lift the seed material form the feeding hopper and delivers it to the top for final release to either seed treater grader or Indented cylinder etc. these are of different height as required and availability of space of Godown.

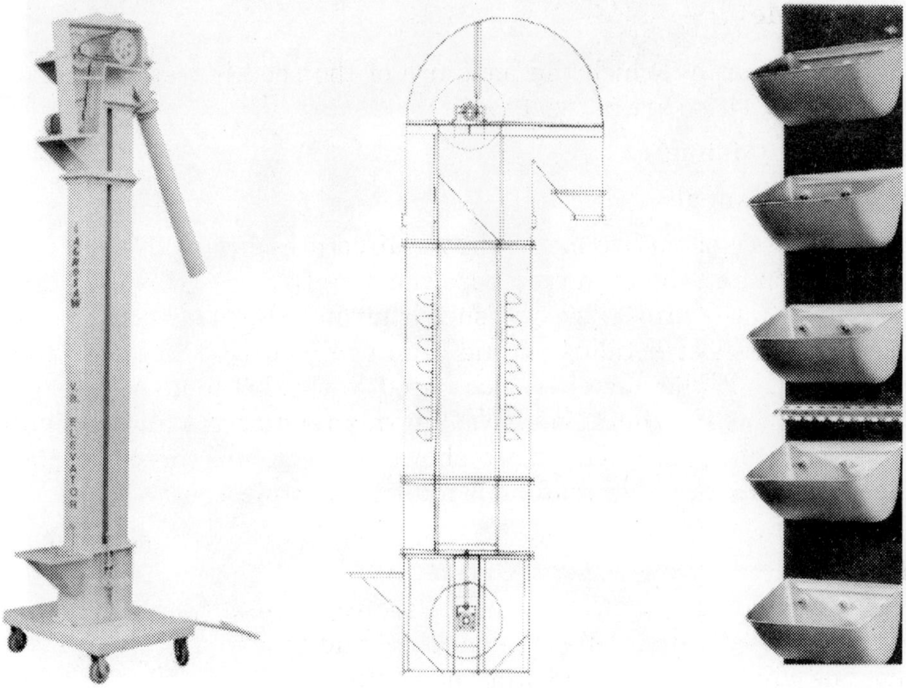

Fig. 3.13. Elevator

Fig. 3.14. Elivation of processing Plant showing Precleaner, Grader & Gravity saprator.

Fig. 3.15. Elivation of G-Nut desheller

Fig. 3.16. Elivation of processing plant.

Chapter 4

Seed Production

Before the matter of production we must know about the Crop seasons there are mainly three crops seasons by which the whole production is dependent.

1. Kharif Season
2. Rabi Season
3. Zaid Season

Kharif Season : It starts from June to October Bajra, Moong, Moth, Cotton etc., are the general crops which are sown in this season.

Rabi Season : Sowing of Rabi crops start from October to April of the next year Wheat, Mustard, Barley, and Gram are the main crops of this season.

Zaid Season : This season starts in between Rabi and Kharif. It generally starts in February to June. It is seen that Moong, Maize gives good yield in Zaid also.

Before going to the production we see the charts for our guideline.

Control on seed Production

It was viewed how the Seed Industry may flourish at its own foot. As they were started under National Seed Project thus Government of India decided to have control over state seed farms and allotment of breeder seed was taken in their hands.

Breeder Seed Indent is sent to Government of India (GOI) by State Corporation and GOI pass the allotment to the concern Breeder for production. The production of Breeder seed is again intimated to GOI by Seed Breeder and GOI allots the seed as Indented as per demand of the states. By the this way the ultimate control over the Seed Industry. Government of India has control all over the Seed Industries. Now we must discuss the type of seed generally used is Seed Industry.

Seed Industry

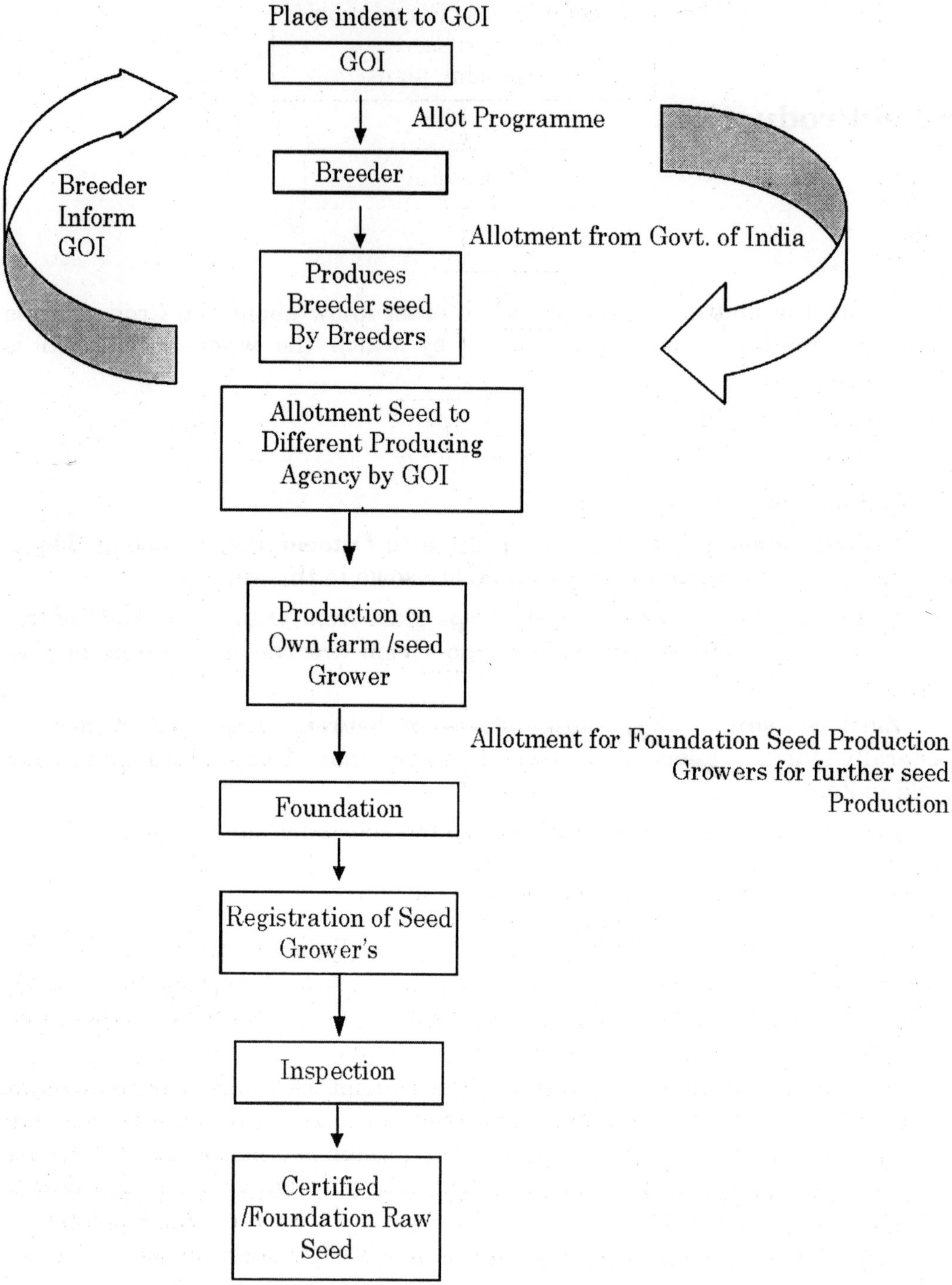

II

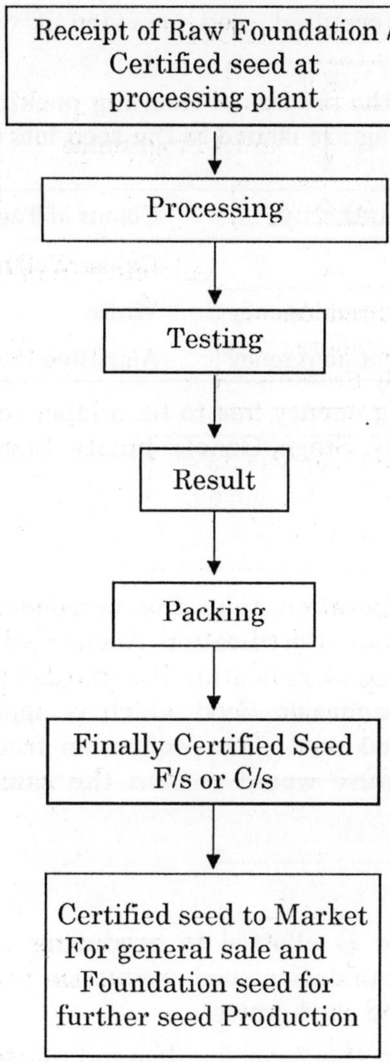

1. Breeder Seed : It is the seed or vegetative propagation directly controlled by the originating Breeder engaged in the Breeding programme. He personally supervises production. The Breeder or the authorized Scientist produces it on the Research station, on directions of GOI as required by state seeds corporation produce its qty. State Seeds Corporation lifts it on demand.

2. Foundation Seed : Progeny of Breeder Seed is Seed is propagated on the specialized farms of Registered Seed grower. This seed is called as Foundations Seed. The production shall be supervised by the Certification Agency. The production is further multiplied by Producing Agency.

3. Certified Seed : The progeny to Foundation Seed in turn certified by State Seed Certification Agency is called Certified Seed. The production is so handled that specific genetic identity and purity can be maintained. Certified seed may be progeny of certified seed provided it could not exceed three generation.

As shown in chart II the process of cleaning packing, Testing takes place in each seed packing the tag are issued to the seed lots as follows: -

Seed	Tag Issuing Authority	Colour of Tag
Breeder	Breeder	Golden Yellow No.356 IS: 5-1978
Foundation	State Certification Agency	White
Certified	State Certification Agency	Ajar Blue IS:

Every seed - producing agency has to fix a label containing the following information : Crop, Variety, Stage, Genetic purity. Size of label must be 15" x 10" colour opeline Green.

PRODUCTION

The State Seed Corporation take the breeder seed and multiply to Foundation Seed. The State Certification Agency will inspect the field in reference to Minimum Seed Certification Standards the production which is received is known as Foundation Seed which is again given to registered grower to produce certified seed. The production from Breeder to certified seed the following procedure would remain the same for seed production Agency.

1. Allotment of Seed

In case Breeder Seed is allotted to producing agency, and producing agency will multiply as Foundation seed production programme on their own farms or through registered seed growers.

Farmer should fill up the form for demand of Breedeer or Foundation seed to registered processing plant. They should register their season wise and pay the following charges for allotment of seed.

1. Charges of seed.
2. Inspection charges per acer.
3. Registration charges (season wise)
4. Other charges as applicable

Note: Private producer can get the foundation seed on demand from other producing agency. They must be in contract with the producing Agency in state or other than state seed corporation for their demand.

2. Sowing of Seed

After allotment of seed it is taken to the producers/ producing agencies field and is sown in the field in prescribed norms: -

1. Seed treatment (with pouch with the seed, if seed not treated)
2. Minimum distance (Row to Row)
3. Depth
4. Manure
5. Isolation Distance (according to self pollinating crop and cross pollinating crop as per MSCS)

Note : Crop wise production can be get from the "Hand book of Agriculture" ICAR publication New Delhi.

3. Intercultural Operation

After Sowing of seed the intercultural operations are conducted by the producer/ Agency or farmer at the field.

Intercultural operation includes timely manure (if required) Irrigation Insect control, timely rouging must be done for assuring non-contaminated seed till the seed is fully ripened for harvesting. Please see that timely Inspection should be conducted by state Certification Agency the required number of Inspections are as follows -

1. Flowering Stage
2. Milking Stage
3. Harvesting Stage

At least two inspection (1) and (3) must be there so as to get the yield potential of the field.

4. Harvesting

The seed is to be harvested after the 3rd inspection. In this estimated yield is received by the certification Agency by the producer. The rejected in Isolation must not be harvested with the crop, which is not taken to processing plant. Seed must be dried properly so that seed must not be damaged by the higher moisture contents and free from insects' damage. If seed is filled at high moisture than insects can damage seed.

5. Filling of bags

The selected harvested crop may be filled in bags tightly and sent to the processing plant by transportation with name, variety stage is marked on the bags.

For those Agency/ Producer Who had their own farm

The agriculture product is produced on any farm the total cost of production can be get through farm management.

Farm Management

Farm Management deals with the organization and operation of a farm with the objectives of maximization profits form the farm business on a continuous basis. Farmer or agency charge methods, price, variability and resources available to him to maximum profit. We can get the minimum production cost of the produce. The base dicissim of farming business are as follows:

i. What to produce or what combination of different enterprise to follow.

ii. How much to produce and what is the most profitable level of production.

iii. What should be size of an individual enterprise that in turn determines the best overall size of the farm business?

iv. What method of production should be used?

v. What and where to market.

For seed industry mentioned above will not applicable. The produce is taken to the seed-processing unit.

The total costing involved in the seed production when divided by actual seed production that unit production cost is get through it. It is therefore the farm budget Performa is recommended.

Ref: Farm Budget Hand book of agriculture

Through above format the total value/amount expended value of by product (HUSK) is deducted than we get the value for the raw seed received it the product is cleaned at the field than the reject value may also be deducted thus the per quintal expenditure value is obtained.

The Unit can only be obtained by using farm budget.

(vi) Transportation : The Raw seed is now transported to the Seed Processing Plant.

Intake of seed

When seed is brought to the seed processing plant the seed is decoded weighed and stacked in the godown. The seed is inspected at the time of intake. The following procedure if maintained then we can get better quality. The inspection at processing plant consists.

a. Physical appearance of seed.

b. Insect infestation of seed.

c. Diseased seed or luster effect

d. Moisture of seed.

e. Expected yield by SSCA must be verified. Otherwise permission must be sought from SSCA. Or verify the contamination if any. Now the seed is ready for cleaning and processing.

Chapter 5

Seed Processing

5.1. INTRODUCTION

The seed Processing is done to up-grade the seed quality by removing the foreign material inert material, undersize seed and weed seed so that the seed can grow properly in the field and a uniformitry is maintained in the field. The field is free from contamination, weeds and free from insect pest. Or The process by which the raw seed is cleaned and the unwanted particles are removed. Finally we get the homogenous seed. Earlier it is shown the flow process of Grader the Seed finally got is free form dust, and unwanted materials.

As per flow chart raw seed is feeded in the hopper of the grader or in the elevator, seed is lifted into the grader. In the grader seed is shorted as per grades by the standard sieves recommended as per MSCS.

Those crops that contain cut seeds we use indented cylinder, the cut seed and weed like Hierankhuri in wheat are removed. For removing weeds in crop like vegetables than gravity separator is used.

The expected percentage fraction of different components of an average seed lots are as follows :

1. Scalped material 1 to 2%
2. Graded Seed 80 to 85%
3. Undersize Seed 10 to 12%
    1. Stone immature seed Including weed seed 1.5 to 3%
    2. Inert material including straw, Dust 2 to 3%
    3. Processing losses including Moisture losses 0.5 to 1.0%

Say 80% Graded

19% sub grade (Undersize)

1% loss

(Loss in the G-Nut varies due to moisture, as moisture loss is higher.1% loss equals to appx 1.200 kg) **(It vary crop wise)**

One should be very particular while handling small grain like Bajra, Mustard, Lucerne, Berseem and vegetable seed etc. Here it is also to be noted that while handling seed like soybean, being very sensitive to mechanical damage.

For Soyabean padded elivator is used or it is feeded by hand. Please note that do not use Cup-Elivator as it will effect germination as its embryo is situated just near the seed coat and is damaged if feeded in the bucket elivator.

Same way every care should be taken after accepting cotton for ginning. After delinting of seed the approximately 60% of total weight is available as Raw seed and the 40% is cotton. This cotton is graded as A, B, C graded.

The percentage recovery of different parts of cotton is seen as follows: -

100 kg Raw Seed cotton

Ginning (at ginning Factory)

65 Kg Seed + 34 Kg Lint + 1 Kg to 1.5 kg loss

Delinting

60 Kg Seed + 3-4 Kg Fuzz + 1-2 Kg Loss

Grading

47-50 kg Seed + 9-12 Kg Undersize and waste + 1 kg loss

Thus total final seed recovery is approx	50-55%
Total lint on an average	35-38%
Fuzz	3-5%
Total loss	3 to 4.5%

The Lower or higher degree of loss may be regularly checked and can be minimised by close monitoring.

Although the above are not proved as universal truth but it is observed in above range. If the higher a degree of gravels weed seed etc. are observed in graded seed than Indented cylinder and gravity separates must be used. Also the wind flow and screen size may adjusted in such a way that in graded seed shall not have seed of size lower than the bottom screen size used beyond 5% (by weight).

Ref: Minimum Seed Certification Standard (MSCS) (1988) Page No. 7 Para XIX

Chapter 6

Certified Seed Packing

When Raw seed is received at seed processing plant the following records are to be maintained by the producing agency.

1. Raw seed Register
2. Grading Register

Raw Seed register contains the following information:-

1. Raw seed Register

Farmers' name	Crop	Variety	Stage	Exp yield	Mois- ture%	No of bags	Qty in qtls	Code No.	Bala- nce	Grading Register page No	Remark
1	2	3	4	5	6	7	8	9	10	11	12

Records to be maintained by State Seed Certification Agency & Producing Agency Both (Grading/Processing Register)

Crop	Vari- ety	Stage	Date of Gra- ding	Lot No. Issued	No. of Bags	Qty Per Bag	Gra- ded Seed	Cumul- ative total	No. of Bags	Under size seed	Sig. SSCA Rep	Sig. Pro- ducing agency
1	2	3	4	5	6	7	8	9	10	11	12	13

Creation of the Seed lots

As per seed act seed lot is denoted as under: -

1. Month & Year of Produce
2. State code
3. Plant code
4. Code of farmer as per grading register serial should be from 01 to.........

E.g. Rabi code : APR 04-20-01-01

Kharif : Oct 04-20-01-01

Means

Month & Year of Produce Apr-2004/Oct04

State code 20 (Rajasthan)

Plant code 01 (Durgapura Jaipur)

Code of farmer as per grading 01 (Rabi/Kharif)

Seed grading record is maintained in Gradind register of Certification Agency and Producing Agency the Graded seed, Undersize, moisture of seed loss etc is recorded.

Seed samples of graded Qty are sent to the state level Authorized seed-testing laboratory for seed testing as per norms. After receiving the testing report seed is packed in the recommended size of packing. Seed report contains Purity/Germination/ Moisture /other crop seed/dead seed etc in the submitted sample.

On the above report under the supervision of State seed certification authorized representative get packed in the required packing size.

Again SSCA has to put the record of STL and the Tags issued to the standard Lot

crop	variety	stage	Certificate No issued/ Date	Lot No	Qty. in lot	Tags issued From To	No. of Tag missing with SR No.	Total Tags issued	Qty. packed	Sign. SSCA Rep	Sign. of Producing agency
1	2	3	4		5	6	7	7	8	9	10

After packing he issue a certificate under section 9 of the SEED ACT to producer agency by SSCA. The information of the seed lot is mentioned i.e. Purity/Germination/ Moisture/other crop seed/dead seed /Number of tags used and qty packed in the lot.

Foundation seed is treated and recommended doze of chemical is used to treat the Standard seed lot by means of seed treater. Seed is dried and then packed in the standard recommended packing. Sec. 9 certificate is also issued to the foundation lots. The original copy is issued to the producer /Producer Agency.

For foundation seed Grow-Out Test is done through SSCA and this seed is packed after the test result of GOT and Seed Testing Laboratory both.

Certified /foundation seed tags give the information of Crop, variety, stage, Certificate No, seed validity etc. By this we can know the viability of the seed lot.

For Foundation seed	White Tag is issued	(Size 15"x7.5" cm)
For Certified seed	Ajar Blue Tag is issued	(Size 15"x7.5" cm)

An Opeline Green label for general information is compulsory on each bag by producing agency. It is the information is put regarding seed lot is confirming the Seed standards as prescribed in MSCS.

The label will contain the following information:

Name of the Producing Agency

Crop	Variety	Germination
Stage		Purity
Date of test		Genetic Purity
Valip upto		(Season Rabi/Kharif/Zaid)

Size of the label will be 15x10" cm.

For the first time seed is tested validity is of nine months. If the seed stored more than nine months than revalidation is done. Seed revalidation is again valid for six Months provided it confirms the standards of seed testing.

Revalidation: When after the season is over and the seed is left over then these carryover lots when used again in the next season the seed are again tested for its conformity of seed standards. This process is called revalidation. The detail of seed lots Crop/Variety/ Qty wise is offered to the State seed certification Agency for the revalidation process. The authorized seed certification officer draws the seed lot samples and sample is sent to State level authorized seed Testing laboratory. The tests are done and the as per recommendations in MSCS. As the test result is received, when lot is found standard the stamp for extension of the validity period is put on each bag and countersigned by the authorized representative on each bag or new tags are issued on request to SSCA. The tests for the revalidation is done for physical purity, germination, and insect infestation only.

Revocation of Certificate : It can be revoked under misrepresentation as to the essential facts or Producer fails to comply with the condition subject to which the Certificate is granted then the certificate can be revoked by SSCA.

Chapter 7

Costing in Seed Industry

INTRODUCTION

Each Industry is based on the cost of Raw material, labour, overheads finally comes to the market for sale with the minimum cost to compete the market.

Costing

According to Carter : Cost accounting is defined as the system of reading accounts of material used and labour employed in the manufacture of certain commodity or a particular job.

According to WWBigg : Cost accounting is the provisions of such analysis and classification of such expenditure as will enable the total cost of any particular unit of production to be ascertain with reasonable degree accuracy and at same time to disclose exactly how much total cost is constituted.

According to Hord J. Wheldon : Costing is classifying recording and appropriate allocation of expenditure for the determination of the cost of products or services and presentation of suitability arranged data for purpose of control and guidance of the management, it includes the ascertainment of the cost of every order, job, contract, process, service or unit as may be appropriate. It deals with the cost of production, selling and distribution.

As indicated above costing deals with the cost of production selling and distribution. Costing in seed Industry is also includes the actual production finally received processed packed and ready for sale at each selling point or distribution point.

There are many types of methods for cost accounting in Industries prevails.

HISTORICAL OR ACTUAL COSTING

In this method costing is done after the completion of the job. The estimated and actual cost difference is thoroughly studied and the bilodge are noted that they are not getting repeated in future.

1. Estimated Costing: It includes the tentative material cost; labour cost etc. It is calculated on the basis of last year's expenditure and difference in cost of any commodity i.e. added in the tender price.

2. Standard Costing : It is calculated on the basis of general output basis i.e. if the particular job is ascertained than the figures are tentatively taken for its minimum standards are fixed, after when actual figures are received than variation is studied why they are differ.

OBJECTS OF COST ACCOUNTING

The product cost is based on the total investments to produce the product ultimate to the consumer on consumer price. To fix the price the cost accountancy is compulsory to get the price of the product. There are some parameters to get the actual price.

COSTING IN SEED INDUSTRY

It is seen that the pooled price is calculated is Seed Industry. The average of all is taken and the price is calculated. It is also seen that these average pooled prices are giving losses in the corporation and it is generally seen in the Government Sectors where the corporation are going in the losses. The following methods can be more beneficial to the corporation where these can be calculated on the following basis.

First we should put the items incurred in the grading and packing we have to breakup into these two features.

1. Grading Cost
2. Packing Cost

The following sheet said to be (AR) Avinash Ramdeo Grading cost calculation sheet and AR (Avinash Ramdeo) packing cost calculation sheet. Which includes the followings. AR-Grading Cost Sheet.

Items

1. Stacking and Weighment
2. Storage/ Fumigation/ Destacking
3. Labour (Grading point)
4. Restacking (graded/undersize)
5. Electric Charges
6. Machine/ Plant Depreciation
7. State Seeds Certification Agency Charges (as Applicable)
   1. Supervision/shift

2. STL Charges
3. Sample forwarding charges by SSCA

AR-Packing Cost Sheet

1. Grading Charges : As per above sheet
2. Labour : Destacking Weighment, Filling bags/ seed treatment (Dry/ Wet Pouch)
3. Cost of chemical for treatment
4. Cost of cloth bag or Jute bags
5. Cost of Lead Seal
6. Cost of Labels
7. Cost of Thread Spool
8. Cost of Sutli (secondary Packing)
9. Cost of 100/60 kg bags (packing cloth bags secondary packing)
10. Miscellaneous (Bin cards, stamps/ Stencil/ repair of weighing machine etc.)

State Seed Certification Charges

1. Tags
2. Supervision/ Shift

The details of grading charges are as follows :

1. Stacking and Weighment : Firstly the Raw Seed received form seed production is weighed and send to processing plant the final budgeting/ costing starts from this point the actual cost of labour rate is taken. Now if there are many seed processing plant than the pooled method is taken.

E.g. Plant	(A) Production Plant A	1500 qtls @ 1.00 =1500
	(B) Production Plant B	500 qtls @ 2.00 =1000
	Pooled prices is Rs. 1.25 per quintal	

2. Storage/ Fumigation/ Destacking:-Now when seed is stored for the turn for grading the Seed is kept pest free. The fumigation is done. For the turn the bag is again taken to the grading shed/machine.

The amount for destacking the bag per qtls. to grading machine. It is likely near Rs. 0.75 per qtls.

Storage: for minimum Rs. 3/ qtls
Fumigation: for once Rs 1.25 per qtls

The total amount may be taken as Rs. 4.50/- (Round off or as per amount paid nett)

3. Labour (grading Point): For labour for Direct packing crops and Indirect packing crops are different.

e.g. Wheat

1. One labour at bag opening point at feeder.
2. One labour at graded seed outlet.
3. One labour at weighing point.

We can cumulate daily Labour cost to Raw Seed handled per day. Or we can call tender in which we bid the labour contractor to give per quintal rate from destacking of raw seed bag to final graded seed. In case of Bajra. i.e. Destacking of seed putting up at grader, after grading of seed graded seed is stacked and also undersize seed is weighed and the final seed rate is given for the whole work. Or the tender price of whole work is taken.

4. Electric Charges : The electric charges are calculated is the following way:-

We can take example for 10 qtls seed

Suppose 10 qtls seed is graded is one hour that

1. Grading Machine (5HP) i.e. 760 x 5 = 3800 watts/ hour
2. Elevator (2) Nos. i.e.1.5 HP x 2 = 1520 watts/hour
3. 2 bulb/ points i.e. 100 watts x 2 = 200 watts
4. Exhaust fan 500 x 2 = 1000 watt
5. Bag closer 150 watt x 1 = 150 watt
6. Others 500 watt = 500 watt
7. Intend Cylinder 760 x 2 = 1520 watt

7690 watt Say 8000 watts used on 10 qtl. Therefore total electricity used is 8 units, for 10 qtls, if the commercial rate is Rs. 6/- this will be calculated on 8 qtl of graded seed received. Same way we can calculate on the basis of total machines installed at the plant and the electric consumption on the total Horsepower consumption (total wattage).

For example

The electricity and other components are calculated on one tone Seed Grader, if grader is of more capacity we had to add the wattage and Raw seed, graded seed qty is taken per hour i.e the total electricity is calculated on graded seed received per hour. If all shown machinery is not installed at your processing plant then total wattage must be deleted from the calculations thus the cost will be reduced.

NB : Only the total installed machinery is to be taken with their wattage only. Example for seed grader 7.50 HP with the accessories shown in sheet.

Crop	Electric Grader	Exhaust Fans	Electric Point	Bag closer etc	Indent cylinder	Scalper Gr. sapretor	Total Watt	Rate 6 Rs per qtl. approximate commercial rate
	7.5 Hp/ hr	2 x 500 W	4 x 100 W	1 x 100 W	1.5 x 760	7.5 hp tot		On Raw seed
Moong	5700	1000	400	100	0	5700	12900	77.40
Moth	5700	1000	400	100	0	5700	12900	77.40
Urd	5700	1000	400	100	0	5700	12900	77.40
Arhar	5700	1000	400	100	0	5700	12900	77.40
Cowpea	5700	1000	400	100	0	5700	12900	77.40
Bajra	5700	1000	400	100	0	5700	12900	77.40
Guar	5700	1000	400	100	0	5700	12900	77.40
Sorghum	5700	1000	400	100	0	5700	12900	77.40
Maize	5700	1000	400	100	0	5700	12900	77.40
Wheat	5700	1000	400	100	1140	5700	14040	84.24
Barley	5700	1000	400	100	1140	5700	14040	84.24
Gram	5700	1000	400	100	0	5700	12900	77.40
Til	5700	1000	400	100	0	5700	12900	77.40
Soybean	5700	1000	400	100	0	5700	12900	77.40
Paddy	5700	1000	400	100	0	5700	12900	77.40
Mustard	5700	1000	400	100	0	5700	12900	77.40
G-Nut	5700	1000	400	100	0	5700	12900	77.40

W = Wattage per hour

Here the total cost of electricity will be taken on the actual graded seed. e.g. if 10 qtl of Moong raw seed is feeded to the machine and 8 qtl seed is get from 10 qtl. Than though the total electricity is consumed on 10 qtl but will be calculated on total output received i.e. 8 qtl and per qtl electricity charge is calculated.

5. Machine/ Plant Depreciation: The machine and plant depreciation is taken as 10% after year of total cost. 10% value is taken and the per quintal seed is brought for ward i.e. total plant/building cost is Rs. 5 lack and 10% value comes to Rs. 50,000. The total seed comes to plant is say 10000 qtls that deprecation comes to Rs. 5/- per qtls seed. The production of seed, more the production lesser the deprecation.

6. State Certification Charges

1. Supervision Charges/ Shift Charge (As applicable according to rates.)
2. Seed testing charges (as applicable)
3. Sample forwarding charges (as applicable)

Here it is taken as if there are small lots of 30-40 qtls than if forwarding charge is Rs. 25/- than per qtls to comes to Rs. 1.25/- we have to make some standard for the calculation to add in the grading cost.

7.5 HOW TO CALCULATE DIFFERENT COST INCURRED PROCESSING MATERIAL

Now it comes to packing cost this Includes the following things.

1. Labour for Packing of Seed: Firstly after getting the result of graded seed is ready for packing. Labour will first de-stake the bag and put up to packing floor than calculate per qtls. of labour charges. The packing cost is calculated on the total packed seed to the amount given to the labours.

2. Cost Chemical treatment: The recommended doze per qtls seed to the purchase cost we can calculate the actual will of chemical per quintal. i.e. **Thiram 75%** Purchase cost 80 Rs. Kg Treatment Rate per qtls is 200 gm/ qtls Rate per quintal is Rs. 16/-.

3. Cost of Jute/ Cloth Bags : Cost of Jute bag is calculated according to packing e.g. Wheat 40 kg. packing, Mustard/ Bajra 3 kg packing, than if purchase cost of jute bags is Rs. 16/- than per qtls. Wheat packing will be 2.5 bag per qtls. E.g. Rs. 40/- per quintal, for Bajra and Mustard required bags are 33.3 bags or Round of 34 bags, If it is Rs. 7/- than the cost of bags is 34 x 7 = 238 Rs.

4. Cost of Lead Seal/ Labels : Again the purchase of lead seal is taken generally the seal weights 2.5 grams each. If lead seal purchase costs is Rs. 50 per kg. than one kg will have 400 seals and each seal will cost 5000/400.

=12.25 paisa per seal

Thus in case of Wheat 2.5 bags x 12.25 = 30.62 or 31 paisa / qtls. In case of Mustard 34 bags x 12.25 paisa = 4.165 or say Rs. 4.17 / qtls.

5. Labels: Now we take the labels value where the per label cost is there, suppose it costs 10 paisa each than incase of Wheat it will 25 paisa/ qtls and in case of Mustard Rs. 3.40 per qtls.

6. Thread Spool: For thread spool we have to pool the value became the actual packing must be calculated per cone. As per experience the following is taken.

1. Cone can stitch (Thread Spool-2)

i. 250 to 275 bags of 40 kg capacity

ii. 350 to 400 bags of 2/3/6 kg capacity

iii. 500-550 1 kg capacity

Add 2 to 5% due to breakage and mishandling etc.

The Cones purchase cost is taken and as per above calculation the stitching of bags costing can be calculated as follows:-

If Cone per bag supplied @ 40 per cone than it comes to 4000/250

16 paisa/ bag approximately.

To stitch 40 kg bag and for quintal ie 16 x 2.5 = 40 paisa per qtls for sticking of 1 qtls seed wheat seed in case of Mustard / Bajra 4000/ 400 = 10 paisa per bag, for Mustard 34 bags of 3 kg are used per qtl packing of mustard thread cost will come to Rs. 3.40 per qtls approximately.

1. Cost of 100/60 kg. Bags (Secondary Packing): The 3 kg packing or smaller are packing is finally done in 100/60 kg. Jute bags. If the packing is in odd number than the packing is done in 75/60 kg and packing in even number, than packing is in 80/60 kg packing. e.g. 2 kg packing of Castor than packing will be done as 80/60 Kg standard packing than 40/30 bags will be filled per 100 kg/ 60 kg. Jute bag than 1.25 or 1.67 bag will be used to pack one quintal of seed.

These days 60 kg packing is recomended.

After purchase cost of 100 Kg jute x 1.25 will give the per quintal packing cost of secondary packing for 80 kg packing and x 1.33 for 75 Kg packing or x1.67 in case of 60 kg packing. All these concepts are to be considered as mentioned above if there is Mustard that 25 bags of 3 kg in one bag than 1.33 bag is required for packing one quintal seed. The purchases cost x 1.33 will give the costing for jute bags per quintal.

Now a days 60 kg packing is in the trend & is likely to come into force by GOI.

8. Miscellaneous : The cost of Bin cards, Stencils, Colour etc. may be fixed in Round off figure.

Tentative cost structure of crops

		Labour	Raw seed		Electric	Exhaust	Electric	Bagcloser	Indent	Scalper	Total	Rate	Rate	Grading	Packing	bags	Rate	Cost	Cost of	Cost	Packing	Cost of	RSSCA	RSSCA	Total
Crop	Weighment	Charges	feeded	Graded	Grader	Fans	Point	etc	cylender	gr. saparetor	watt	6 Rs per qtl	per qtl	Cost	size	required	jute/ cloth	of	Lead seal	of	80kg/ 1.25	Secondary	Tags	packing/	likely
	Raw seed	grading	per/ Hour	qty/ Qtl	7.5 Hp/hr	2X 500	4X 100w	1 X 100W	1.5x 760	7.5 hp tot		on Raw seed	on graded	A	Kg		Bag	Bags	0.15	labels	75 kg/ 1.33	Packing	round	supervision	
Moong	2.50	7.00	8	6.4	5700	1000	400	100	0	5700	12900	77.4	12.09	21.59	4	25	5.2	130	3.75	3.75	30	37.5	25	3	224.59
Mooth	2.50	7.00	8	6.4	5700	1000	400	100	0	5700	12900	77.4	12.09	21.59	4	25	5.2	130	3.75	3.75	30	37.5	25	3	224.59
Urd	2.50	7.00	8	6.4	5700	1000	400	100	0	5700	12900	77.4	12.09	21.59	4	25	5.2	130	3.75	3.75	30	37.5	25	3	224.59
Arhar	2.50	7.00	8	6.4	5700	1000	400	100	0	5700	12900	77.4	12.09	21.59	4	25	5.2	130	3.75	3.75	30	37.5	25	3	224.59
Cowpea	2.50	7.00	8	6.4	5700	1000	400	100	0	5700	12900	77.4	12.09	21.59	4	25	5.2	130	3.75	3.75	30	37.5	25	3	224.59
Bajra	2.50	7.00	8	6.4	5700	1000	400	100	0	5700	12900	77.4	12.09	21.59	3	34	5.2	176.8	5.10	5.10	30	39.9	34	3	285.49
Guar	2.50	7.00	8	6.4	5700	1000	400	100	0	5700	12900	77.4	12.09	21.59	4	25	5.2	130	3.75	3.75	30	37.5	25	3	224.59
Sorghum	2.50	7.00	8	6.4	5700	1000	400	100	0	5700	12900	77.4	12.09	21.59	4	25	5.2	130	3.75	3.75	30	37.5	25	3	224.59
Maize	2.50	7.00	8	6.4	5700	1000	400	100	0	5700	12900	77.4	12.09	21.59	5	20	5.2	104	3.00	3.00	30	37.5	20	3	192.09
Wheat	2.50	7.00	10	8	5700	1000	400	100	1140	5700	14040	84.24	10.53	20.03	40	2.5	16	40	0.38	0.38	0	0	4	3	67.79
Barley	2.50	7.00	8	6.4	5700	1000	400	100	1140	5700	14040	84.24	13.16	22.66	30	2.5	20	50	0.38	0.38	0	0	4	3	80.42
Gram	2.50	7.00	10	8	5700	1000	400	100	0	5700	12900	77.4	9.68	19.18	30	3.33	20	66.6	0.50	0.50	0	0	4	3	93.78
Til	2.50	7.00	8	6.4	5700	1000	400	100	0	5700	12900	77.4	12.09	21.59	1	100	4.5	450	15.00	15.00	30	37.5	100	3	642.09
Paddy	2.50	7.00	8	6.4	5700	1000	400	100	0	5700	12900	77.4	12.09	21.59	30	3.33	20	66.6	0.50	0.50	0	0	4	3	96.19
Mustard	2.50	7.00	8	6.4	5700	1000	400	100	0	5700	12900	77.4	12.09	21.59	3	34	5.2	176.8	5.10	5.10	30	39.9	34	3	285.49
G-Nut	2.50	25.00	6	4.8	5700	1000	400	100	0	5700	12900	77.4	16.13	43.63	30	3.33	30	99.9	0.50	0.50	0	0	4	3	151.53

Foundation seed		Treater		Rate		qty/hour	elictric	cost of	qty	rate	Total cost
		Watt	1.5 X 760	per	charge	treated	charge	chemical	used /qtl	per qtl	Tentative
		1.5 Hp	Kilo watt	unit			per qtl	Thirum/kg	kg		
Moong	224.59	1140	1.14	6	6.84	8	0.855	125	0.3	37.5	262.95
Mooth	224.59	1140	1.14	6	6.84	8	0.855	125	0.3	37.5	262.95
Urd	224.59	1140	1.14	6	6.84	8	0.855	125	0.3	37.5	262.95
Arhar	224.59	1140	1.14	6	6.84	8	0.855	125	0.3	37.5	262.95
Cowpea	224.59	1140	1.14	6	6.84	8	0.855	125	0.3	37.5	262.95
Bajra	285.49	1140	1.14	6	6.84	8	0.855	125	0.3	37.5	323.85
Guar	224.59	1140	1.14	6	6.84	8	0.855	125	0.3	37.5	262.95
Sorghum	224.59	1140	1.14	6	6.84	8	0.855	125	0.3	37.5	262.95
Maize	192.09	1140	1.14	6	6.84	8	0.855	125	0.3	37.5	230.45
Wheat	67.79	1140	1.14	6	6.84	8	0.855	125	0.3	37.5	106.15
Barley	80.42	1140	1.14	6	6.84	8	0.855	125	0.3	37.5	118.78
Gram	93.78	1140	1.14	6	6.84	8	0.855	125	0.3	37.5	132.14
Til	642.09	1140	1.14	6	6.84	8	0.855	125	0.3	37.5	680.45
Paddy	96.19	1140	1.14	6	6.84	8	0.855	125	0.3	37.5	134.55
Mustard	285.49							125	0.3	37.5	322.99
G-Nut	151.53	1140	1.14		0	8	0	125	0.3	37.5	189.03

9. SSCA Charges : SSCA will take its charges for packing and tagging the rates may be calculated accordingly.

e.g. 1 tag cost 1 Rs. in case of Mustard 33.3 tags x 1 will come to Rs. 33.30 and incase of Wheat 2.5 x 1 Rs = Rs. 2.50 per quintal of seed.

As the charges of SSCA is modified time to time then the modification may be done accordingly in respect of shift/tags or revalidation charges etc.

COST STRUCTURE

1. Procurement price : we have to decide the procurement policy to procure the seed or the procurement price of the seed.
2. Processing and packing expenses as per above calculations.
3. Storage charges on carryover seed. Pool the price
4. Oveheads
5. Business overheads or percentage for the whole seller and retailer
6. Transportation overhead.
7. Packing loss/transportation loss etc
8. Finally retail price.

COST STRUCTURE FOR COTTON SEED

In case of cotton we can calculate the cost incurred in packing of cotton seed per quintal.

1. Raw cotton procurement price
2. Weighment of raw cotton
3. Ginning charges
4. Delented charges
5. Delented bales pressing charges
6. Delented seed weighment and bagging charges
7. Transportation charges : Raw delented seed from ginning factory to Seed Processing plant
8. All the other charges applicable as per raw seed of other commodity i.e. processing and packing expenses.
9. Loadind and unloading transportation charges of delented bales to the processing plant.
10. Seed certification charges (at each stage of ginning delinting and grading of seed)

11. Other charges like insurance, Mandi tax added.
12. Now we get the total quantity of graded seed, undersize, numbers of delented cotton bales. The bales. undersize included fuzz are to be auctioned and the amount received is to be deducted from the actual cost so we get the final cost on packed seed received
13. Now again same cost structure mentioned above like other crops.

Chapter 8

Seed Storage

INTRODUCTION

The purpose of Seed Storage is to protect the seed from up and dawn of the nature i.e. the physiology of seed is effected by various storage conditions.

Seed when reaches the physiological maturity on plant the seed possess its maximum viability and vigor. The Disease, Insect damage, Mineral deficiency etc. reduces maximum vigor obtained at maturity.

From the maturity stage to seed again planted in the field in next season the seed is to be stored. The character of seed is that it never stopped the aging process thus it also dies in respect of germination. It will never germinates. If it is stored many years through it can only be used as grain only for feed/food and fodder.

WHY SEED IS STORED

1. Sowing for next season
2. Maintain the Buffer Stock
3. Maintain the Parental lines of Hybrid
4. Conserve the germplasm

The science cannot stop aging and loss of germination but he can retard it by power storage conditions. Storage life of seed (Paddy, Wheat, Maize, Sorghum, Millets) stored all temperature not exceed, 90° Farahanite at different Seed Moisture contents.

Seed Moisture Content	Storage Life
11-13%	½ Years
10-12%	1 Years
9-11%	1 Years
8-10%	2 Years

8-9% can be packed in sealed container.

For general and most vegetable seeds subtract 3% from the seed moisture content for cereal to determine there after storage life at 90° F. The relation of Seed Moisture to other factors that effects the survival of the Seed in storage.

Seed Moisture level at which factors occur. Which are eterious seed in storage. It depends upon the "T" the temperature.

T Decline in vigor during storage

This "T" depends upon

1. Relative Humidity 2. Temprature
3. Pathogen/Insect/Pest 4. Chemical-Pesticide

SEED MOISTURE

40-60%	Germination Occur
18-20%	Heating may Occur
12-14%	Molds grow on and in seed
8-9%	Insect active and Multiply
5-10%	Sealed storage is safe.

See storage and packing Application for India. If is seen that at 12% moisture no fungi has ever been found at 75°F.

James F. Hassangton
Johnson E. Douglas

If the moisture is in higher side than the moisture is to be lowered.

Mechanism of drying : Seed generally gain moisture in the outer surface of the cotyledon it can be regulated by Air-flow by increasing of temperature or relative humidity. Dehumidification of air is done by

(i) Using chemical desiccant

(ii) Refrigeration

(iii) *Using of Silica Gel:* It is the neutral substance it does not react with the seed material and it regulates the seed material in the store/storage Tanks. This silica gel is rechargeable by heating on the Iron tubs

Optimum Temperature for the seed for seed drying

Seed moisture	Drying temp. (°C)
>18%	32

10-18%	38
< 10%	43

Optimum Moisture of Seed

Cereals and pulses	12-13%
Vegetables	8-9%
Moisture pervious pack	5-6%

Thumb rule for storage

1. For every decrease of 1% in seed moisture the life of seed is doubled.
2. For every decrease of 10°F in storage temperature the life of seed is doubled.
3. Temperature + Humidity must be less than 100.

Good seed storage in an important phase of processing and is essential for successful seed marketing. For better seed storage of seed in seed processing plant the following measures are to be taken which will give safe seed storage.

(i) Water drainage system: Water drainage system should be proper in processing plant. One should see that no leakage in walls is there otherwise flows of water in seed godown will generate/ increase moisture in seed & seed will destroy.

(ii) Sanitation: Processing plant should be cleaned in intervals regularly so as to remove dust and other particles from floor and walls. Floors and walls are to be cleaned properly holes are the main source of insect and pests. Daily sweeper should clean the floors of godown.

(iii) Use of insecticide and Pesticide: Regular use of insecticide and pesticide will control insect and pest in the processing/ other godowns.

(iv) Area Calculations : One should calculate the actual area, which is required for i.e. According to seed production and seed to be processed. One bag area approx. 43½ x 26½ with or it will cover area up to 6 sq. ft. For stack side area where on cover or used for spray/ fumigant, it will came to 0.6-0.66 sq. M/ton than required area should be calculated on basis of production programme.

Crop	**Height**
Wheat, Bailey, Oat, Maize	
Jowar, Bajra, Pulses	20 Nos.
Paddy (dhan)	18 Nos.
Soyabean	6-8nos

If condition storage spare is evaluated place storage need in advance, preferably at the time of hazard. Stacking must be done on the insect free wooden crates. Stack must not be more then 30' x 20' and at least 2½ feet must be left between wall and stack. If cerates are not available than seed polythene 300 gaze should be used under the raw/graded seed.

(v) Moisture in the Seed: Moisture in the seed plays an important role in seed storage as the moisture is increased the seed Industry must be alert. The insect/pest get good atmosphere to develop. Themselves the seed moisture content and seed storability charts is as follows:

(vi) Seed Moisture Content and Seed storability

The role of moisture in the life of a seed is as follow:-

Seed Moisture Content	Effect on Seed
35-80%	Moisture content of developing seed, seed not mature enough for harvest.
18-40%	Seed physiologically mature, reparatory to field deterioration, heating occurs if seed bulked without adequate ventilation, molds and insects very active, seed susceptible to mechanical damage in harvesting and handling.
13-18%	Respiratory rate still high, can get heating at higher levels, molds and insects can be damaging, seed resistant to mechanical damage.
10-13%	Seed storage reasonably well for 6 to 18 months in open storage in temperate climates, insects can still be a problem in susceptible seed, seed susceptible to mechanical damage.
8-10%	Seed sufficiently dry for 1 to 3 years open storage in temperate climates very little insect activity seed very susceptible to mechanical damage.
4-8%	Safe moisture content for sealed storage.
0-4%	Extreme desiccation can be damaging to seed handedness develops in some kinds of seed.
33-60%	Seed germinate when they imbibe water to these levels.

EFFECT OF TEMPERATURE ON SUSCEPTIBILITY

Temperature is the main factor affecting toxicity of fumigants to insects. The uptake of fumigant is linked to the rate of respiration of the insect. Temperature is particularly important in phosphine fumigation while relative humidity has a relatively minor impact on the susceptibility of insects.

The rate of deterioration of crop seed in storage increases as seed moisture content increases.

Mature seed are hygroscopic their moisture content vary with the relative humidity of the atmosphere, although not as quickly as changes occur in atmospheric relative humidity. Moisture absorption or desorption of seed is a relatively slow process.

Equilibrium moisture content varies among different kinds of seed. In general, the equilibrium moisture content of only seed is lower than that of starchy seed at the same relative humidity and temperature.

While both seed moisture content and temperature are important factors in seed storage, moisture content has greater direct influence own seed longevity. Dry seed can store well at temperatures up to 90°F.

A One (1) percent decrease in moisture content nearly doubles storage potential of seed and a ten (10) degree decrease in temperature nearly doubles storage potential of seed.

As the temperature and humidity ratio or both added to touch 100 than it is a good atmosphere to develop insect/ pest. If comes near 100 that is an alarm for Industry to be alert in contest with the Insect multiplications if it is seen in godowns. We should go for spray and fumigation as required. The fumigation a spray must be regularly done in 10-15 days as follows: one must not do spray in air tight godowns.

Safe Seed Storage : Good seed storage is an important phase of processing and is essential to successful seed marketing. Proper storage preserves seed viability from harvest to sale, and protects the produces, the processor, and the user.

Storage must be provided for seed from the time of harvest until planting time. Storage must also be provided for seed that may be carried over until the planting season in the next or a subsequent year. There must also be long-term storage for lots of foundation and breeder seed to preserve valuable germplasm. Seed that is in transit from place to place is actually to the finally stored.

STORAGE PROBLEMS ARISE BECAUSE

1. Low quality seed are placed in storage.
2. Inadequately dried seed are stored.
3. Storage godown is not properly constructed or proper storage conditions are not maintained.
4. Seed are stored for too long period.

SEED QUALITY IS NOT IMPROVED BY STORAGE

Seed coming out of storage are never higher in quality than were when placed in storage. Seed deterioration is inevitable and irreversible low quality seed cannot be made into high-quality seed even through the best storage conditions, nor can germination be held at high levels for many years.

Seed that have become to deteriorate because of adverse pre-harvest environment, harvesting damage, heat damage during drying, or storage at high moisture content will lose germination rapidly in storage.

Practical procedures for improving seed storability are:-

1. Carry over only high quality seed.
2. If conditioned storage space is limited, use it for storage of good seed, and plan storage well in advance, preferably at the time of harvest.

Temperature : Temperature plays an important role in the life of seed, although it is not as important as moisture. Within the normal range, there is increased biological activity of seed, insects and molds as temperature rise.

In bulk stored seed at high moisture at high moisture content, metabolic activity of seed and associated microorganisms, produces heat, which raise the temperature of the seed mass. The increase in temperature accelerates biological activity, which causes further heat production and completely kill the seed embryo.

Note : Spray is not to be done in the airtight godowns only fumigation is to be done.

FACTORS AFFECTING SEEDS IN STORE

The most important feature of preserving seeds for long periods is based on maintaining viability and other attributes of quality seed above prescribed minimum standards. Therefore, it is important to understand various factors that directly or indirectly influence seed during storage.

A. A biotic factors

Temperature: This is one of the factor that influence metabolic activity of all biological material. At higher (< 20ºC) temperature, most of the microbes, mites and insects, multiply and infest seed affecting its health. Thermal requirements of different groups of pests are as follows:

Bio-agents	Temperature (ºC)			Moisture content (%) minimum
	Minimum	Optimum	Maximum	
Insects	15-20	27-37	33-38	9.0-11
Mites	5-12	19-31	32-36	-
Fungi	2-5	19-31	34-40	15.4-16.8
Microbes	5-6	28-35	50-58	-

Humidity: Like temperature, humidity is also an important ecological factor which influences seed longevity and pest activity. Most of the seed under tropical environment deteriorates quickly due to high humidity coupled with temperature.

Seed moisture: Moisture content in seed is another factor that directly affects seed deterioration and pest multiplication. Moisture content below 10% in seed is considered good for storability. For keeping seeds in moisture-proof packets, the moisture content requirement of different important storage insects is as follows:

Storage insects	Moisture content (°C)	
	Minimum	Maximum
S. oryzae	9.5-11.0	14.0-14.7
R. dominica	9.0-10.0	11.0-14.0
T. granarium	0.0-1.9	11.5
T. castaneum	10.0	11.5-16.0
C. cephalonica	9.0	15.0-20.0
C. cautella	10.0	16.0

Two principles in safe seed storage are: (1) the percentage relative humidity plus temperature (°F) should not exceed 100, and (2) with each 1% reduction in moisture content (Range 5-14% m.c.) or 5°C drop in temperature (Range 0-50°C temperature), the storability of seed without the loss of germination gets doubled.

Oxygen availability: The role of oxygen in respiration is well known. In an airtight container, the oxygen content of the air is consumed in respiration both by seed and by the insects, if any. It is estimates that oxygen consumption by insects is 130,000 times that of seeds. Hence, oxygen content is depleted faster in insects infested lot than in insect free lots. Each insect species requires a particular minimum concentration of oxygen for its survival and multiplication, and they cannot survive when the oxygen content goes below that level. The minimum concentration of oxygen required by different stages of two species of storage pests has been determined at IARI. This clearly indicates the requirement of oxygen for survival and multiplication of storage insect pests.

Insect species	State	Oxygen content (%)	Reference
Trogoderma granarium	Egg.	16.77	Girish, GK, 1964
	First instars larvae	5.35	Ph.D. Thesis of IARI,
	Full grown larvae	1.08	New Delhi
	Adults	3.39	
Tribolium castaneum	Full grown larvae	6.37	Sinha, BP, 1965
	Adults	7.24	M.Sc. Thesis of IARI

INSECT PESTS OF STORED SEEDS

1. Rice Weevil, *Sitophilus oryzae* Linn.

This species is a serious pest of paddy and other cereals. It prefers temperature but humid climate and breeds wherever it finds undisturbed condition of storage. It is tiny weevil, generally reddish brown but sometimes dark brown in color, with a snout-like head. The size of adult largely depends on the size of seed/grain and prevailing weather conditions.

Both adults and larvae attack the whole seed. They feed voraciously rendering seed unfit for even human consumption. In case, of heavy infestation seeds become a lump of broken debris. Damage is maximum between August to November, less in February to March and Least in winter (December-January) and during hot dry summer.

The female makes a small excavation in the part of the grain where it lays around 250 translucent and white eggs. These hatch in about four to six days depending upon the weather. The tiny young grub is white with a yellow brown head. It bores into seed and feeds on starchy contents hollowing it out to leave only the shell. Usually, only one grub is found inside a seed, though sometime there may be two or three depending on the seed size. The grub stage lasts for 20 to 30 days. After passing through a one or two day's prep pal state, it pupates inside the hollow grain for three to six days depending temperature. The adult weevil emerges from the infested seed and starts a new generation. The average period from egg to adult is one month. The number of generations in a year depends on prevailing temperature and humidity condition. Usually, there are five generations in a year but seven or eight are also on record.

2. Lesser Grain Borer, *Rhyzopertha dominica* Feb.

This is a major pest of all cereals. A badly infested seed is hollowed out until a thin shell remains. Up to four beetles may be found in a single large seed such as maize. Its attack on wheat in the field has also been reported.

The female generally lays eggs on the seed near the embryo end, which is comparatively soft and allows the young larvae to enter easily. A single female may lay 300-500 eggs. The duration of eggs stage is five to six days in summer and longer in winter.

The newly hatched larva is quite active and compodiform in shape. It harrows at once into the seed or crawls actively feeding on loose starch material. There are four or five moulds and the full-grown larva is dirty white with light brown head and curved abdomen. An average larval period is 44 days. The pre-pupal and pupal period last for 7 to 8 days. Total life cycle from egg to adult takes about two months. There are generally five generations in a year.

3. Khapra beetle, *Trogoderma granarium* Everts.

This is a serious pest of whole cereal seed, especially wheat. Generally, its infestation occurs on the top layer of seed because the insect fails to penetrate beyond a certain depth. In a massive infestation, it can affect the whole lot. As rule, it first attacks the embryo point and then the whole seed.

Adults appear late in the spring and eggs are laid among the grains. A single female lays around 125 eggs in her lifetime. These hatch in about 6 to 16 days depending on the prevailing temperature and humidity. In humid atmosphere the incubation period is 5 to 6 days only.

The larva is brown-white in color. The body is covered with a bundle of reddish brown movable hairs on the posterior segments forming a sort of tail. The larval period varies considerably. Under favorable conditions it is completed in about 50 days with four moulds. But under unfavorable conditions, the larval period is prolonged from 200 days to four years. During winter or when it is without food, the larva remains inactive and lives in cracks and crevices.

Pupal stage last for 6 to 17 days. Adult lives for 10 to 30 days after reaching maturity.

4. Pulse Beetles, *Callosobruchus chinensis* Linn. & *C. maculatus* Fab.

Pulse beetles are the most destructive pests of stored pulses, namely moonbeam, cowpea, pigeon pea, chickpea etc. infestation starts from the field when the crop is maturing. They also infest the seed godown. The pest develops during storage and is detected when adult beetles emerge from seeds. Subsequently, and adult beetle lays eggs on seeds causing further loss to the seed quality. Although this insect remains active throughout the year, its infestation is Maximum from July to September causing 40-50% loss.

The female lays eggs either on maturing pods in the field or on the seeds in the godown. Eggs are glued on the surface of seed. A female lays around 50 to 100 eggs in her lifetime. Under favorable environmental conditions, eggs hatch in four to five days. The grub bores into grain immediately after hatching and spends its life inside the seed. It moults four times and completely develops in two to three weeks. It then transforms into a pupa. The adult normally emerges in four days but in winter it may take up to 28 days. Complete life cycle from egg to adult takes around 25 to 30 days in summer and 50 to 60 days in winter. The adult normally survives for about 10 days. In a year, 7 to 8 generations are recorded.

5. Grain Moth, *Sitotroga cerealella* Oliv.

Like pulse beetle, this species also infests seed crops in the field. Being a lepidopteron, only the larval stage is injurious to seed. Adult only laid eggs on crops in the field or on seeds or seed bags. After emerging from the egg, a larva enters the seed through a minute hole and feeds on it. It completes its life cycle inside the seed before emerging as adult (moth). In this way the larva completely destroys the seed leaving behind the shell. Its infestation is generally detected only after the emergence of moths.

The female lays around 400 eggs. The larva emerges in about four days during summer and seven days in winter.

It feeds inside the seed and spends its life there, molting five times and developing in 14 to 20 days. It pupates inside the seed in a cocoon. The moth emerges in 7 to 10 days.

PREPARATION OF SEED STORE/SHED

Before arrival of the new harvest, clean all processing and storage structures thoroughly followed by disinfestations with residual sprays or insecticides at least four weeks in advance. If there is a history of insect infestation, a second spray should be given a week before arrival of the new harvest. Treatment is best carried out when the godowns are empty.

PRE-STORAGE RESIDUAL SPRAY OF INSECTICIDE

Surface 100 Sq.m.	Volume of Water (I)	Malathion 50 EC (ml)	Fenitrothion 50 EC (ml)	Methyl Primiphos 50 EC (ml)
Metal	2.5	25-50	25	50
Concrete	5.0	50-100	50	100
Brick	7.0	70-140	70	140

Apply insecticide solution as given in the above table using a knapsack sprayer, and volume of solution sufficient to wet all floor and wall surfaces and to fill cracks and crevices. Also, treat trucks and trailers, processing plant building and other structures where insects can hide. Before loading the sprayer, care should be taken to mix the insecticide thoroughly with the correct amount of water.

Soak the old gunny bags in hot water (>50°C) for 15 minutes and dry them, or treat with Malathion @ 10 ml/liter of water per 20 Sq.m. bag surface (both sides), or fumigate them with aluminum phosphide. 1 tablet of 3 gm./m^3 of space under airtight conditions with an exposure period of 5-7 days to kill all hiding and surviving stages of insects.

Store clean and dry seeds. The moisture content of seed should preferably be below 10%. At this level of moisture content most of the insect species do not breed or multiply.

Do not store the new harvest with carried with carried over seed stocks. Also ensure that the new harvest does not carry any insect infestation. If live insects are detected in the seed lot it must be fumigated before moving it in the godowns.

Insect infestation normally occurs after threshing and winnowing. Some insects lay their eggs of maturing crops, especially in pulses in the field. They multiply during pre-processing period. Such infestation is detected when adult insects emerge from seeds and considerable damage is already done. In such cases, fumigate the seed lot under cover with aluminum phosphate (Celphos/Quickphos) @ 2 tablets of 3 g each/tone of material with an exposure period of 5-7 days. This kind of pre-processing care is essential for pulse seeds because pulse beetle (bruchid) infestation comes from field and increases rapidly. Ensure moisture content below 12% before fumigation.

Storing pulse seeds in bags with close weaving and tight stitching can also check pulse beetle attack. Bags should be machine stitched with cotton thread because hand stitching with jute thread (Sutli) leaves free spaces at the top and in folds. The pulse beetle lays its eggs through such folds on the top of the bags.

Direct admixing of insecticides provides long-term protection to seeds, but treated seeds must be kept in a cool dry place. Also the moisture content of seed should not be above 10%. Otherwise insecticides impair viability and vigor of the treated seeds. In no case should insecticide treated seeds be disposed of as food or feed. The insecticides used for admixing are listed in the table below.

Insecticide	Dose Per / MT Tone	Remark
Malathion 5% dust D	200 g	Premium Grade
Methyl Primiphos 2% D	200 g	
Malathion 50 EC	60 ml in 5000 ml water	
Methyl Primiphos 50 EC	12 ml in 5000 ml water	
Deltamethrin 2.8 EC	40 ml in 5000 ml water	

CARE DURING STORAGE

After proper stacking, treat the left free surface (alley ways), stack, doorways etc. with any one of the insecticides given in the following table. This provides protection to new produce for about three months. After fumigation, surface treatment also restricts re-infestation.

Insecticide and its dosage for surface treatment

Insecticide	Dosage per metric tone of seed
Malathion 5% D	2.00 g
Methyl Primiphos 2% D	1.25 g
Malathion 50 EC	50 ml per 5 liter of water
Fenestration 50 EC	50 ml per 5 liter of water
Methyl Primiphos 50 EC	50 ml per 5 liter of water

When monitoring of insects is not being done by other means as described above, inspection of seed lots at an interval of two (rainy and spring seasons) to four (summer and winter season) weeks is most important for the management of insect pests in the seed godowns. If live insects or other evidence of infestation are detected, fumigate the seed lot under air-tight conditions either in the room or under cover made of low density polyethylene sheet of 1000 gauge or any other suitable material.

FUMIGATION UNDER SHEET OR COVER

Sheet or cover fumigation is commonly practiced in sheds. This provides better gas-tight conditions and control of insects than space fumigation. The stacks of bagged seeds are covered with a gas proof sheet made of high density 1000 gauge polyethylene sheet with perfect (double) sealing at joints or a rubber sheet braving provision for dispensing phosphine tablets. The most important points to remember are that the size of sheet should be sufficient the to cover the stack properly with margins at the ground level for proper sealing. The dose should be worked out according to the quantity of seed rather than the size of cover. The cover should also be protected from rodents, which can damage the seal causing loss of gas.

Minimum durations of fumigation with phosphine under different temperature regime

Temperature (°C)	Duration of fumigation, days after application of fumigant
< 12	Do not use phosphine
12-20	6-7
22-25	5-6
26-30	4-5
>30	4

REFERENCES

1. Seed technology - *RL Agarwal*
2. Safe Seed Storage of Seeds Insect Pest Management - *S.N. Sinha*
3. Lectures attended at NSC on seed Storage - *Malavika Dadlani* (2005)
4. Viability of seeds - *E H Roberts* (1972)
5. An introduction to Seed Technology — *J.R. Thompson*

Chapter 9

Seed Quality Concepts and Seed Certification

QUALITY CONCEPTS

Though the importance of good seed was realized with the inception of agriculture, however, the concept of seed quality on scientific lines, gained importance during 1869, when professor Friedrich Nobbe in Germany advocated that the seeds must be tested before sowing. This gave births of the establishment of seed testing laboratories in European countries, USA and Canada during late eighties and early nineties. The establishment and development of seed testing laboratories in the developed world generated tremendous impact in the seed trade and the development of seed testing procedures. The international seed Testing Association (ISTA) was established in 1921, and the first set of International rules of seeds testing were framed and published by ISTA during 1931, since then, the seed-testing programme has been expanded throughout the world.

Seed quality is the sum of all those attributes that differentiate the seed from the grain. The important seed quality attributes are: Genetic purity, physical purity, germination moisture, health and vigor. In addition to above quality seed should be of uniform size and should possess good luster.

MOISTURE

The life processes of the seeds revolve around the seed moisture content. The moisture content of the seeds should not be too high or too low. Moist seeds are difficult to thresh and will deteriorate at the faster rate. Similarly, if the seeds are too dry, it may shatter in the field and thereby reduces the yields. Seed moisture content should be brought to the desirable level, through drying (nature or artificial) enabling to preserve their viability in storage and avoiding the spoilage by insects and pathogens. The moisture content also determines the choice of the packing material for marketing or storage purposes.

The moisture content of the seeds varies according to its chemical composition; it is less in case of oily seeds white it is high in case of those seeds where reserved food material is either predominantly carbohydrate or proteins.

SEED HEALTH

Seed health is also important seed quality attribute, especially under tropical and sub-tropical conditions. Quality seed should be free from pests and pathogens. Infection of seed by pests and pathogens may affect germination and vigor and thereby reduces the yields. It has been found that a large number of micro-organism causes considerable damage to the seed quality either by discoloration of the seeds, reducing the seed size and weight or causing the seedling or adult plant diseases. Infected seed lots are also responsible for the spread of the disease in the disease free areas. It is, therefore, essential to test the seed lots to measure be status of its health in order to avoid problems in seed production and quality control. In seed certification scheme, the tolerance limits for the designated seed borne diseases of the vegetable and field crops have provided to avoid economic losses. In addition, seed treatment with fungicide or insecticide and fumigation is a regular practice for controlling the disease organisms and pests.

SEED VIGOR

The performance potential of a seed lot with reference to field establishment is very such dependent on the capacity of the seed to germinate, emerge and establish under sub-optimal field conditions. This performance potential or "hidden stamina" of the seed makes it fit to perform well upon sowing. This is called seed vigor. It has been found that the field establishment of two seed lots of the same cultivar having same germination percentage may not necessarily be the same upon planting. One seed lot may give rise to good emergence and establishment while the other may fail to do so in the same field. This differential behavior in the performance of the two seed lots is mainly due to their vigor status. One seed lot may be more vigorous in comparison to the other seed lot. Vigorous seed usually provide higher yields in comparison to less vigorous seed. According, the measurement of seed vigor is of considerable importance in assessing the seed quality. In addition, the storability of the seed is influenced by its vigor status. Vigorous seed lots will store well for longer periods without appreciable loss in the germination in comparison to the less vigorous seed, which deteriorate at a faster rate.

GENETIC PURITY

This is the most important seed quality attribute. The genetic purity of the seed lot governs the yield potential for he variety. It is, therefore, very important that the genetic (cultivars) purity status of the seed lot should be high. In a seed certification scheme, the genetic purity standards have been prescribed. The breeder seed should be of the highest genetic purity. In the foundation and certified seed classes, the genetic purity standards are lower to maintenance breeding programme. The responsibility of the maintaining the genetic purity of breeder seed, lies with the plant breeder who has evolved the variety, while the genetic purity of foundation and certified seed classes is ensured through seed certification.

PHYSICAL PURITY

The physical purity of the seed should also be high. The seed, which is obtained after harvesting and threshing the seed lot, is not fit for immediate sowing of planting purposes. It contain admixtures, such as soil, stone pieces, flower stalks, broken seeds and pieces of leaves, barks, chaff and other foreign material together with seeds of other crops and weeds. These admixtures makes the harvested seed unfit for sowing or storage. The occurrence of undesirable as these may reduce the yielding ability of the seed lot. During storage, the extraneous matter present in the seed lot serves as he places for harboring the spores of saprophytic fungi, which may cause the loss of seed viability. Therefore, the harvested seed must be cleaned and processed to remove the admixture.

The pure seed, weed seed, other crop seeds and inert matter contents judge the physical purity status of the seed lots. The pure content in a seed lot should be high and those of other crop, weed seed and inert matter should be minimum for obvious reasons.

GERMINATION

Germination capacity of the quality seed lot should be high for obtaining the desired crop stand in the field. Using seeds of low germination will reduce the field establishment or stand and thus the yields will be lowered. Seed germination is affected by a variety of factors, which are imposed to the seed during its formation, maturation, ripening, such as, infection with the pests and pathogen. Sub optimal conditions of nutrients and water supply and untimely rains or frost at the maturity stage. In addition, post harvest operations and handling of the seed lots during marketing of distribution are also responsible for affecting the seed quality.

SEED QUALITY CONTROL

Quality control is an important component of the programme. In fact, the essence of any seed programme lies in the quality control. A seed programme without the provisions of regulating the seed quality control measures may collapse. Seed quality control is a system which ensures to government the quality of the seed through checks, certification and official regulations (legislation), for seed production, it is necessary that the quality of the seed at maximum possible standards. In other words, it may be inferred that the seed handled within the seed programme should be monitored to ensure that its quality is maintained during the production processes in such a way that the seed below certain acceptable standards is rejected.

There are two aspects of quality control. Firstly, the genetic purity of the seed is maintained during the production and marketing operations. Secondly, it should be ascertained that the seed lot itself is of adequate quality. It should be free from weeds, other crop seeds, extraneous material, and disease organisms and possesses high germination capacity and optimum moisture to avoid its deterioration during storage and marketing.

QUALITY MAINTENANCE

The responsibility of maintaining the quality of the seed lies with the producer (public or private) who is producing and marketing the seed. Seed production is a highly specialized and systematic job. It encompasses the production of nucleus seed, breeder seed, and foundation, certified and labeled seed classes. The process occurs in a systematic manner. The breeder seed is produced from the nucleus seed, foundation seed from the breeder seed or foundation stage I and the certified seed from foundation seed or certified state I classes. If there is some missing links in this chain, quality maintenance programme suffers and consequently the quality assurance and quality control components will get the setbacks. It is therefore, imperative that adequate care needs to be exercised in the quality maintenance or internal quality control programme. It, this is achieved adequately well, the problems in the quality control will be minimized and the assurance of seed quality provided by the producer will not face any problem during marketing or distribution.

The first and foremost aspect of quality maintenance is the quality of the breeder seed with special reference to the genetic purity. The breeder seed should be of the highest genetic purity. The genetic purity of breeder seed should be above the genetic purity status of the foundation seed class. In India, the breeder seed does not fails under the PUT view of seed certification. However, its quality must be monitored by the breeder himself or by a join team or any other authority, which has been identified in the rules and regulation. In addition, it is genetic contamination during harvesting, threshing, drying, clearing, and grading and packing operations of the breeder of seed.

In case of vegetable seed crops, seed certification is opted mainly by the public organizations, whereas huge quantities of vegetable seed is being marketed throughout the world by the private seed companies who seldom opt for certification. It would, therefore, be essential that the private seed companies should have their own research and developmental wing for ensuring the quality of their produce so as to serve the farmers in a better way. If they do not achieve this, they will loose faith of the farming community and they will be out from the seed business within no time. To avoid the problems in the seed trade, it is the responsibility of the government to enforce certain measures for controlling the seed quality.

QUALITY ASSURANCE

This is the responsibility of the seed corporation/organization and the seed companies to provide assurance about the quality of the seed offered by them for marketing or distribution. There would not be any problem to the seed producing organizations in providing the assurance of the seed quality, if they are conscious about the maintenance of the quality of their produce. In the absence of the system of quality maintenance, it would be rather impossible to provide the assurance of the seed quality. Without observing the quality maintenance, if quality assurance is provided by an organization in the form of a declaration or a label containing the details of the physical purity and germination of seed, the organization may face heavy penalties under the provisions of the seed legislation of seed Act of else will loose the faith of the consumer. Accordingly, it would be desirable on the part of an organization or seed company involved in the see trade that before providing the assurance relating to the identify of the variety and the quality of the seed offered by them, they must ensure to get the seed tested in the seed testing laboratory.

SYSTEM OF QUALITY CONTROL

Seed quality control in itself is a system, which ensures the interest of the farmers and to avoid hazards in the crop production. It is the responsibility of the government to enforce the measures for controlling the quality of the seeds being marketed in the country. This is usually accomplished through the legislation in the form of a law or Act of parliament. Essentially, there are two components in a seed quality control system namely, the seed Certification and Labeling. Seed certification may not be compulsory but labeling is usually a compulsory provision of any see legislation. To accomplish the task of seed legislation or seed Act, framing the rules and regulations pertaining the legislation and their scope is the prerequisite. In addition, certain basic infrastructure facilities are also required. This includes, the notification of the seed standards, kind or varieties to be governed by the legislation, establishment of seed Certification Authority or Agency. Seed Testing Laboratories, processing plants and notification or authorization of seed inspectors and seed Analysts.

The following are the essential components of a seed quality control programme:

1. Quality control of Breeder Seed.
2. Seed Certification.
3. Seed Legislation.
4. Seed Standards.
5. Seed Testing.

In India, seed quality control programme was initiated during 1963 with the establishment of the National Seeds Corporation, which was the single agency for production, certification, and marketing of the seed throughout the country, however, at a later stage, during 1970's and 1980's State Seed Certification Agencies were established in most of the stages, which are responsible to carry forward the seed certification programme in the respective state.

Seeds Act was passed in India during 1966 and amended in 1974. The promulgation of the Act was made effective during 1969; Indian Seeds Act is essentially "truthful" labeling act. The seed certification in India is voluntary. At present the seeds Act as well as seed Certification in India is applicable to the "Notified" kind/varieties. For the regulation of the seeds Act and the Seed Certification Programme in the country, two separate sets of the Standards namely, Minimum Labeling Standards and Minimum Seed Certification Standards were prescribed during 1970, which have now been revised. The statutory provisions of the Indian seeds Act have got strong linkages with the various components of the seed programme, such as variety release and notification, seed production seed certification and seed marketing.

It would be desirable that initially the seed quality control measures should not be very rigid. Otherwise the seed programme will suffer. However, when the seed certification is becoming more popular and adequate infra-structural facilities have been developed the quality control measures should be applied rigidly at a later stage.

FACTORS AFFECTING THE SEED QUALITY

Seed quality is influenced by a variety of factors imposed to the seed during its formation, development, maturation, growth, harvesting of extraction, during, cleaning, grading, packing, storage and marketing. The factors which governs the seed quality can be broadly grouped into 4 categories, namely:

1. Genetical Factors
2. Ecological Factors

3. Agronomic Packages and Production Technology
4. Harvesting and Post-harvest Handling

GENETICAL FACTORS

Genetic purity or "cultivars purity" may decline as a result of genetic drift occurring during the seed production of a particular variety. The varietal characteristics may change during the production stages due to the combined effect of ecological and other factors. This is, especially true in situations where same sources of the seeds is continuously used over the years. In addition, chromosomal aberrations and mutations resulted during the unfavorable storage may cause the genetic deterioration of the variety.

ECOLOGICAL FACTORS

The growth of the plant and the seed quality are strongly influenced not only by the genetic factors alone but also by the environment prevailing during the seed formation, maturation and growth. The important ecological factors which governs the seed quality include: soil adaptability, wind velocity and rains, light intensities and temperature conditions, insect activities during flowering, pests and pathogens.

These factors may favorably or unfavorably affect the seed quality depending upon the situation.

AGRONOMIC PACKAGES AND PRODUCTION TECHNOLOGY

Application of the prescribed agronomical packages is of utmost importance for raising a good seed crop. The recommended practices may slightly vary according to the location or the season. A personal experience, in this regard is very important in order to suitably modify a particular cultural practice. The cultural practices required for raising the seed crop are different to those of raising the general crop. Timely operations during the crop management is essential aspect for ensuring a rich harvest and to produce the quality seed of desired genetic purity and to avoid deterioration of seed quality by insects, pest and pathogens and the occurrence of undesirable weeds and other crop seeds. Seed source, isolation and plant protection, land preparation, time method of sowing, weed control, rouging of off types and diseases plants, irrigation and nutrition and control of pests and pathogens contributes in affecting the seed quality.

HARVEST AND POST-HARVEST HANDLING

Supervised harvesting and after care of post harvesting handling of the seed crop is an essential component for obtaining the quality seed. This is an integral part of a seed certification programme. The time, energy and money spend in growing genetically pure and disease-free seed crop may go waste, if harvested seed crop id left to the vagaries of the weather. Seed quality is very much affected during threshing or extraction of the seed, processing-drying, cleaning and grading, seed treatments, seed packing, labeling and sealing, storage, fumigation and transpiration of seed lots if adequate technology and care is not provided to the harvested produce.

Choice of appropriate technology for the abovementioned post harvest operations will very according to the location and the kind of the seed. However, the basic principles for these operations are the same throughout. For example, it is established fact that threshing should not be done, if the seeds are too moist or dry. At high initial moisture content, if threshing is attempted, it would be difficult to thresh the seed, or the seeds will be injured. At such moisture levels (18% of above), "brushing" injuries may occur. It would be difficult to find out the "bruised" seed with the naked eye. The immediate impact of "bruising" injuries may not be serious however; there would be sudden decline in viability and vigor in the storage. The "bruise" areas respire heavily and are the centers for infection by the microorganisms. This combined effect of high respiration rate and infection with pathogen will impair the seed viability within no time. Similarly, harvesting or threshing the seed at a very low moisture levels (below 9%) will induce the fracturing or cracking injuries, thereby reducing the seed quality. For the maintenance of seed quality (germination and vigor), it is very essential to dry the seed properly. The choices of drying method (natural or artificial) depend upon the kind of seed and the locality. The safe levels of moisture content for seed packing and storage have also being investigated or important vegetables and field crops. Seeds must be dried properly to the required moisture levels before packing or storing them. Adequate supervision during all the post harvest operations is also required to avoid contamination of the seed lots of different varieties on the threshing floor or in the processing plant. It must be ensured that the labeling of the seed lots has been made correctly and in proper manner.

The state level seed certification Agencies were established to ensure then genetic/ physical / physiological quality and seed health at field level, so as to ensure the genetic identity and Genetic purity varietal character of the crop and variety prescribed by the Govt. of India as per seed Act 1966.

The general Seed Certification Standards are applicable to all crops, which are eligible for certification, and with field and standards for the individual Crops shall constitute the Minimum Seed Certification Standards. It includes all seeds as used in these standards. Seed certification is the process by which the notified crop and variety seed is identified. The genetic /physical/physiological quality and seed health is maintained at field level.

The state level seed certification Agencies also maintain the followings:-

1. List of source of Breeder and Foundation Seed by the Central Seed Certification Board. Certification shall be conducted by Certification Agency the crop and variety notified under section 8 of the Seeds Act.
2. Outline the procedure for submission of application for growing, harvesting, processing, labeling, and tagging of seed indented for seed certification.
3. Verify upon receipt of an application for certification that the variety is liable for Certification.
4. Undertake Inspection of seeds fields, Seed processing plants and seed lots in accordance with guidelines of Central Seed Certification Board.
5. Regulates the processing of seed at seed processing plants.
6. Arrange for analysis of seed samples draw from the seed lots produced as per the procedure to verify their conformity to, the prescribed standards.
7. Ensure the seed lot must confirms the Standards of Central Seed Certification Board.
8. Grant Certificate Tags, to the lots as per the prescribed procedure.

ELIGIBILITY OF VARIETIES FOR CERTIFICATION

Seed of only those varieties, which are notified under section 5 of the Seed Act 1966, are eligible for certification.

PHASES OF SEED CERTIFICATION

Seed Certification consists the following phases:

1. Verification of the seed source and Class of Seed requisite for further seed multiplication.
2. Receipt of scrutiny of the application.
3. Field inspection to verify conformity to the prescribed standards of the MSCS and the gazette allotment at time to time by Government of India.
4. Supervision of post harvest stages including processing and finally packing of the seed lots.
5. Seed sampling and seed analysis including Genetic Purity, seed testing and seed health in order to verify the conformity to the prescribed seed standards by the authorized seed testing Laboratory
6. Grant of Certificate and Certification Tags and bags are sealed for conformity of the of the seed standards.

HOW TO OFFER SEED PROGRAMME

A. Seed producing agency must take license from the licensing Authority Deptt. of Agriculture of the State and permission to marketing seed in the State.

B. Seed producing agency must register the Processing Plant at Seed Certification Agency of the State, which will give the plant code to the Seed Processing Plant.

C. This Code will be the permanent Code. This will be renewal at each year by the Agency. This code will be used to each lot produced under that Seed Processing Plant.

Refer Certified seed packing chapter 6 allotments of seed lots Nos.

Source of seed used for seed production must be subject to:

1. The verification by seed certification by the seed certification Agency i.e. offered list must contain the source of verification. Breeder/Foundation Seed tags /Certificate must be verified by the Seed Certification Authority.
2. Application must be submitted with the prescribed form No. 1 (This form will supply all ware about of the seed source, seed stage, Farmer's field description, Seed taken for seed multiplication and parental lines if any used.
3. Seed producing Agency will advance for the seed certification with the registration fees, Inspection Fees etc as applicable.
4. Seed, which is offered for production. Lot number and validity period, Seed purchase records (source etc.) is being submitted to the seed certification Agency for further seed multiplication.
5. Seed at the most can be produced at the maximum four stages. After that purity is generally not said to be maintained or seed is not liable for seed certification. These stages are as follows:-

Breeder Seed: It is the vegetative propagating material directly controlled by the originating or sponsoring Plant Breeder that provides the source for the initial and recurring increase of Foundation Seed. Breeder Seed be so pure that in the progeny *i.e.* Foundation and Certified Seed shall meet the Standards of Genetic Purity. The other quality factors : Purity, Inert Matter, Germination etc. shall be indicated on the Label on the actual basis.

Foundation Seed: This is the category of seed, which is certified by the seed Certification Agency Notified under Section 8 of the seed Act, 1966. Seed Certification Agency must be recognized by the Central Government through Notification in the Official Gazette.

(A) Breeder Seed

↓

Foundation Seed I

↓

Certified Seed I

↓

Certified Seed II

Certified Foundation Seed stage - I is the progeny of Breeder seed.

1. This Foundation seed must clearly trace to Breeder Seed. It is produced directly by the Breeder Seed.

Certified Foundation Seed-II

1. Certified Foundation Seed produced from Foundation Seed – I is called Foundation Seed – II stage. It is not be further used as Foundation Seed it is only used as Certified Seed-I. Certified Tags are of White Colour in both the stages i.e. II and I. Minimum Certification is same for both the stages (Chart B).

(B) Breeder Seed

↓

Foundation Seed I

↓

Foundation Seed II

↓

Certified seed I

NB: Certified Foundation Seed stage-I is only adopted when Certified Foundation Seed it is expressly felt by the certification Agency that Breeder Seed is in short supply.

Certified Seed-I seed is the progeny of Foundation Seed-I or it may be of Foundation Seed stage-II.

NB: Production of Certified Seed must not exceed total Four stages of the certification i.e. progeny of breeder Seed. After this stage Genetic purity will not significantly alerted.

Same way if Breeder Seed is not in sufficient quantity and seed is already produced as certified seed than certification can go to the further multiplication as certified seed-III (mentioned in Chart - A).

At the time of inspection the following isolation distance is checked to maintain the genetic purity otherwise the field is rejected. The isolation distance of the general crops is as follows:

Kharif

M - male

F - female Seed

Isolation Distance and Nos. of Inspection for different crops

Crop	Seed Rate Per Hacter (Kg) M x F	Isolation Distance in Meter		Number of Inspection
		Foundation	Certified	
Bajra				
Composite	4	400	200	4
Hybrid	1.5Mx3 F	1000	1000	4
Jowar				
Composite	15	200	100	4
Hybrid	3.5M x 7 F	400	400	4
Maize				
Composite	20-25	400	200	4
Hybrid	7m x 14F	600	300	4
Paddy	20-25	3	3	2
Cotton				
Desi	12-15	50	30	2
Hybrid	16-20			
Guar	15-20	10	5	2
Arhar	15-20	10	5	2
Moong	12-15	10	5	2
Urd	12-15	10	5	2
Moth	10-15	10	5	2
Cowpea	15-20	10	5	2
G-Nut				
Bunch (Kernel)	100	3	3	2
Spread (Kernel)	60-80			
TIL	2.5	100	50	3
Soybean	80-100	3	3	2
Castor	12-15	600-1000	300	2/4

Rabi Crops

Crop	Seed rate Per Hater (Kg) MxF	Isolation Distance In Meter		Number of Inspection
		Foundation	Certified	
Wheat	100	3	3	2
Barley	100	3	3	2
Gram	70	10	5	2
Mustard	5	50	25	3
Pea	75	10	5	2
Berseem	20	400	100	2

By maintaining the isolation distance Seed Certification Agency maintain the genetic purity of the foundation and Certified Seed. If the Isolation distance is not maintained by the Seed grower than the part of the field that come under the certification will be rejected will be mentioned as LFR (liable for rejection).

The seed-producing Agency should also be advised to conduct quality control inspection to avoid mixing of seed, which is rejected by the seed certification Agency.

Chapter - 10

Seed Sampling and Testing

INTRODUCTION

The main features of seed Industry is seed testing. The samples are drawn randomly from the graded bags. It is done because only by seed testing we can judge weather the produced seed is up to the mark of certification or not as mentioned in the guidelines and Seed standards of MSCS. While grading of seed, the graded, bags are filled and while weighing from each bag little sample is drawn this pooled sample is mixed and from that samples are taken and send for seed testing. The two samples are taken in each lot. The first samples is coded and sent to Authorized State Level seed testing laboratory. The second samples is preserved as guard sample which is used if the dispute arises then it is analyze it is kept for two years then it is liable to right off. Now the seed is sent to the state level seed-testing laboratory for seed testing as per norms. After receiving the testing report seed is packed in the recommended size of packing.

A sample is obtained from the seed lot by taking small portions at random from different position of the lot and combining them.

SEED LOT

It is a specified quantity of seed, physically identifiable, in respect of which a seed test certificate can be issued.

Primary Sample : Small portion taken from one point in the lot.

Composite Sample: Is formed by combining and mixing all primary samples taken from the lot.

Submitted Sample: Sample submitted to seed testing laboratory. The size of the submitted sample is specified in the Seed Testing Rules.

Working Sample : Is a sub sample taken from Submitted Sample in the laboratory quality test is done.

Under seed law enforcement programme *only trained and experienced* officials are authorize to undertake sampling and he has to:

(i) Give notice to such intention to the person from whom he intends to take sample.

(ii) Take three representative samples in the prescribed manner and mark and seal.

- One sample to be delivered to the person from whom it has been take
- Second to be sent for analysis to the seed analyst of the area.
- Third to be retained for any legal proceedings.

(iii) Sampler must verify the information provided on the label as per the requirements of the Seed Act:

(iv) Following information should be checked on label

  (i) Kind

  (ii) Variety

  (iii) Lot Number

  (iv) Date of Test

  (v) Seller's/Producer name and address

  (vi) At least two persons should be present and obtain the signature of both the witnesses on form VIII of the Seed Rules.

In case cf certified lots sampler should check information on seed certification tag:

- Name and Address of certification agency
- Kind and Variety
- Lot No.
- Name and Address of certified seed producer
- Date of issue of the certificate and its validity
- Class and Designation of seed
- Period during which the seed shall be used for sowing i.e the validity period of the seed lot.

The seed lot should be so arranged that each individual or part of the lot is conveniently accessible.

Upon the request of the sampler, the owner shall provide full information regarding bulking and mixing of the lot. When there is definite evidence of heterogeneity sampling shall be refused. If the nature of the presentation of the seed lot or container makes it impossible to adequately apply these procedures, then the sampling shall not be undertaken and alternative presentation of the seed lot should be sought. The size of the seed lot should not exceed to the maximum size as prescribed in the rules subject to 5% of tolerance.

MINIMUM SAMPLING INTENSITY FOR SEED LOTS

Seed Lot in kg	When seed in bulk
Up to 500	At least 5 primary samples
501-3,000	one primary sample for each 300 kg but not less than 5
3,001-20,000	one primary sample for each 500 kg but not less than 10
20,001 or more	one primary sample for each 700 kg but not less than 40

Minimum Sampling Intensity

For seed lots in containers

Containers	No. of Primary Samples
1 – 4	3 samples from each container.
5 – 8	2 samples from each container
9 – 15	1 samples from each container
16 – 30	15 samples in total
31 – 59	20 samples in total
60 or more	30 samples in total

For seed lots in containers smaller than 15 kg capacity, containers shall be combined to sampling units not exceeding 100 kg and sampling units shall be regarding as containers in the sampling scheme above.

It consist of hollow tube inside, closely fitting outer sleeve has solid pointed end. The tube and sleeve have open slots in their walls So that when the tube is turned until the slots in tube and sleeve are in line. Seed can flow into the cavity of the tube and when the tube is given half turn the opening are close.

EQUIPMENTS FOR SAMPLING

Using the hand triers or probes can do effective seed sampling

(a) Stick trier Diameter 12.7 mm and Length 762 mm having 9 slots

It consist of hollow tube inside, closely fitting outer sleeve has solid pointed end. The tube and sleeve have open slots in their walls. So that when the tube is turned until the slots in tube and sleeve are in line. Seed can flow into the cavity of the tube, and when the tube is given half turn the opening are close.

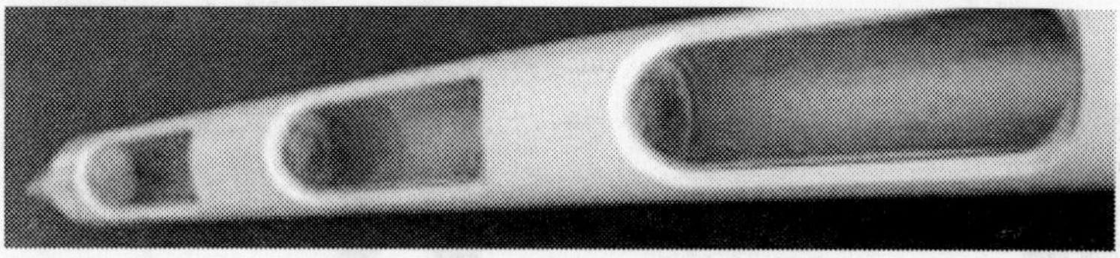

Fig. Stick trier

(b) Bin Trier

It consist of hollow tube inside, closely fitting outer sleeve has solid pointed end. The tube and sleeve have open slots in their walls So that when the tube is turned until the slots in tube and sleeve are in line. Seed can flow into the cavity of the tube, and when the tube is given half turn the opening are close.

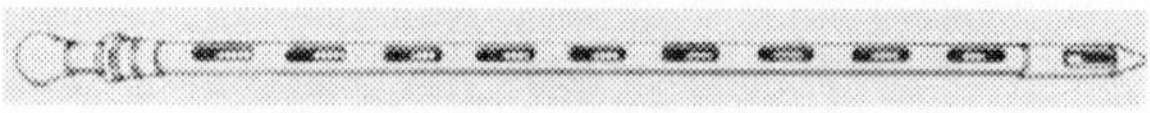

Large trier for sampling from heaps

Nobe triers

Bag trier for sampling small seeds

Bucket or pan for collecting sample

Cloth bag for holding samples
Individual trier

Alkathene bag to Hold 150 Gms of seed

Check lot & count bags

Take sample at random

Empty trier & repeat as required

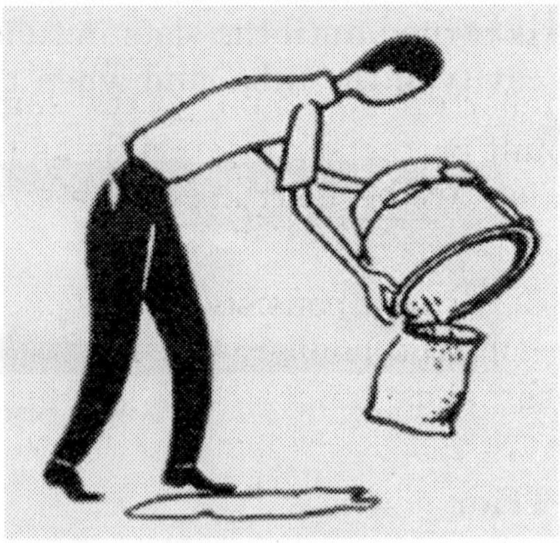

Empty sample into bag

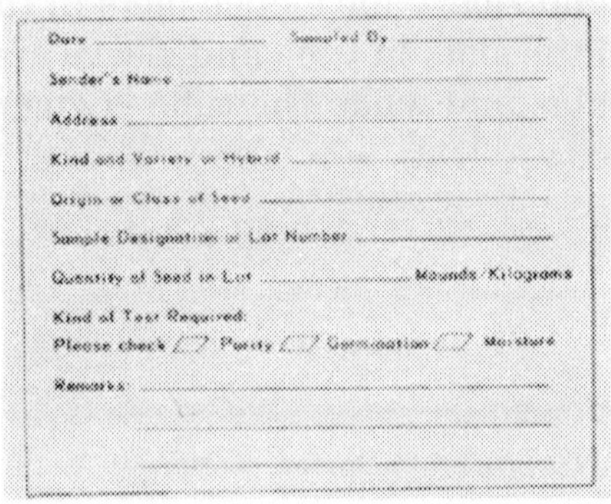

Date .... Sampled By ....

Sender's Name ....

Address ....

Kind and Variety or Hybrid ....

Origin or Class of Seed ....

Sample Designation or Lot Number ....

Quantity of Seed in Lot .... Maunds/Kilograms

Kind of Test Required:

Please check [ ] Purity [ ] Germination [ ] Moisture

Remarks: ....

Label on which detailed information about seed lot to be put inside bag

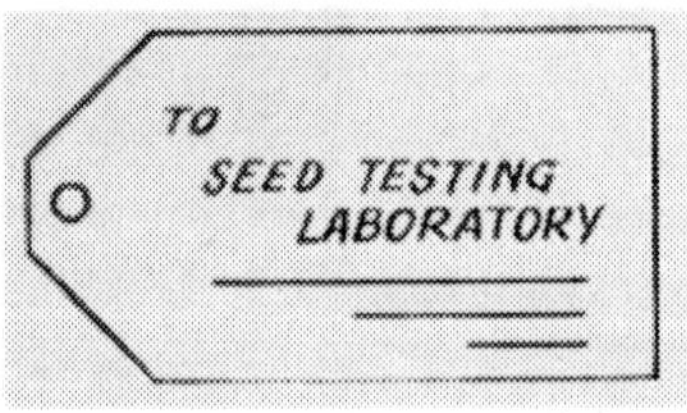

Methods for obtaining working samples

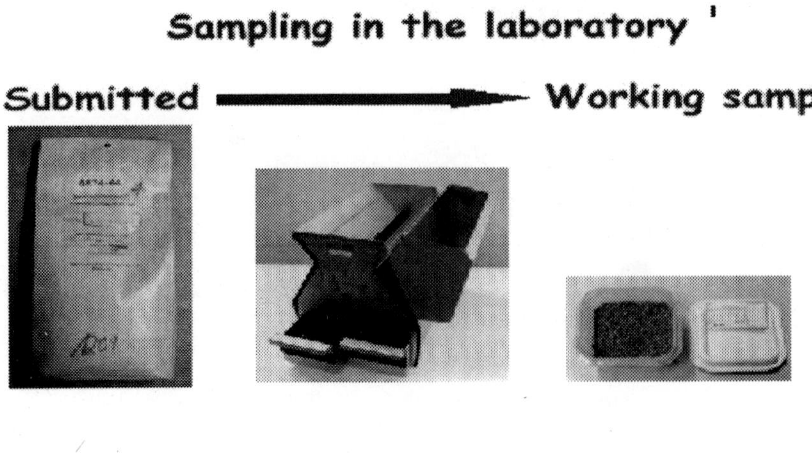

(c) Soil divider

A simpler divider, built on the same principle as the conical a simpler divider, built on the same principle as the conical divider, is the soil divider. The channels are here arranged in a straight row instead of a circle as in the conical divider.

The soil divider consists of a hopper with attached channels or ducts, a frame to hold the hopper, two receiving pans and a pouring pan. Divider, built on the same principle as the conical divider, is the soil divider. The channels are here arranged in the soil divider. schematic drawing of the construction principle of the soil divider. 12.7 mm wide 18 channels

B. soil divider without a pouring aid.

C. the collection pans have a bended edge that can be hang on the bar above the channels.

The following dimensions have been found suitable. Ducts or channels, 12.7 mm wide, lead from the hopper to the collecting pans. There are 18 of these channels, alternate ones leading to opposite sides.

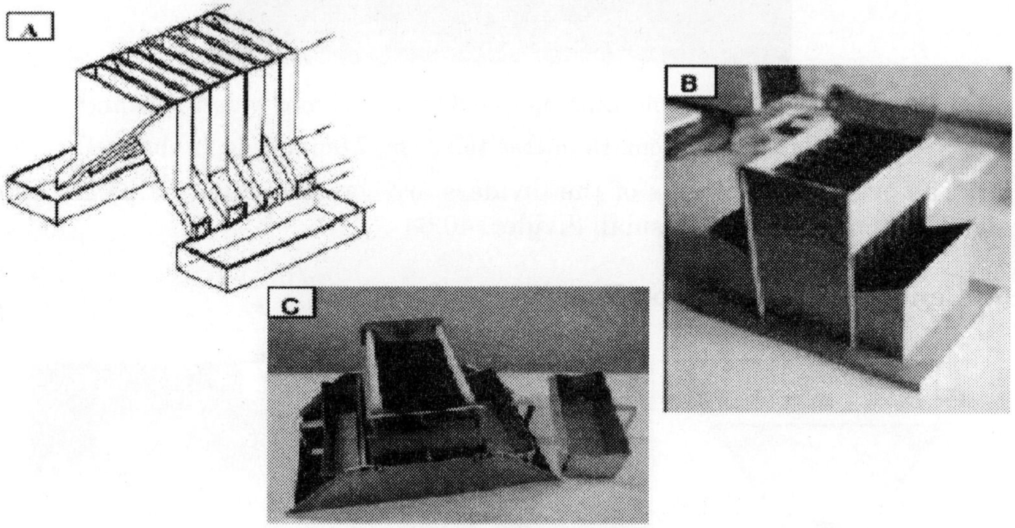

The maximum dimensions are: 35.6 cm long; 25.4 cm wide; and 27.9 cm high.

In using the divider the seed is scattered fairly evenly in a pouring pan the length of the hopper and poured in at approximately equal rates along the entire length of the hopper.

This divider is suitable for large-seeded and chaffy species, but suitable types for small seeded species can also be made.

(d) Conical divider or Boerner type seed divider

Suitable dimensions

Large divider, designed for large seeds and grains, there are 19 channels and 19 spaces, each 25.4 mm wide. Small divider designed for small free-flowing seeds there are 22 channels and 22 spaces, each 7.9 mm wide.

(a) Height 40.64 cm; Diameter 36.83 cm; 25.4 mm wide 19 channel

(b) Height 81.28 cm; Diameter 15.24 cm; 7.9mm wide 22 channel

The overall dimensions of the dividers are large divider, 81.28 cm high and 36.83 cm in diameter; small divider, 40.64

(e) Centrifugal Divider or Gamet Type Divider

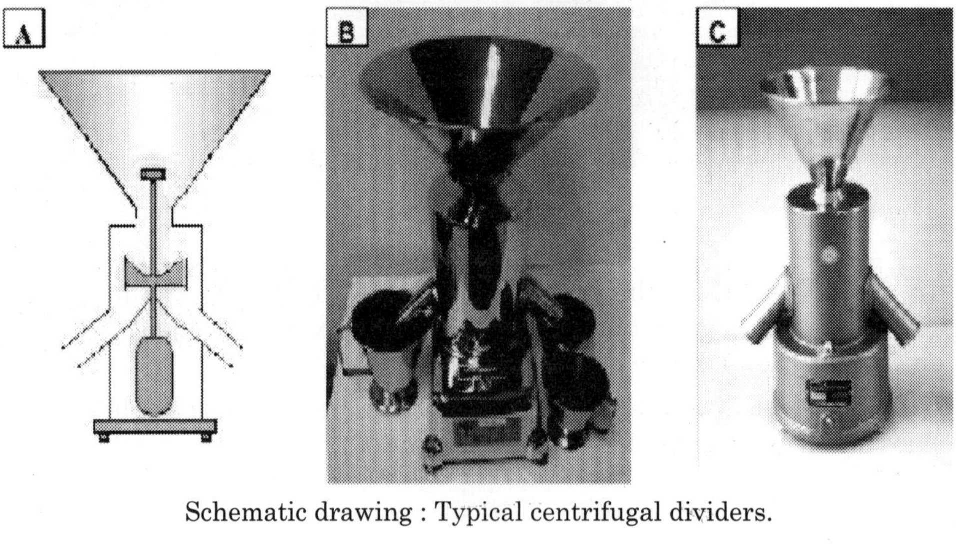

Schematic drawing : Typical centrifugal dividers.

(f) Centrifugal divider

The centrifugal divider (Gamet type) makes use of centrifugal force to mix and scatter the seeds over the dividing surface. In this divider the seed flows downward through a hopper onto a shallow rubber cup or spinner. Upon rotation of the spinner by an electric motor the seeds are thrown out by centrifugal force and fall downward. The circle or area where the seeds fall is equally divided into two parts by a stationary baffle so that approximately half the seeds fall in one spout and half in the other spout.

The centrifugal divider tends to give variable results when not carefully operated.

To avoid variability in the results the divider is operated

(i) Leveled by means of the adjustable feet.

(ii) Divider and four containers are checked for cleanliness.

(iii) A container is placed under each spout.

(iv) The whole sample is fed into the hopper; when filling the hopper, seed must always be poured centrally.

(v) The spinner is operated and the seed passes into the containers.

(vi) Full containers are replaced by empty containers. The contents of the two full containers are fed into the hopper together, the seed being allowed to blend as it flows in. The spinner is operated.

(vii) The procedure described in

(viii) above is repeated at least once more.

(g) Spoon method

Useful for samples of a single small-seeded species and for sample reduction for moisture determination or seed health testing.

A tray, a spatula and a spoon with a straight edge are required. After preliminary mixing, pour the seed evenly over the tray; do not shake the tray thereafter.

With the spoon in one hand, the spatula in the other, and using both, remove small portions of seed from not less than five random places. Sufficient portions of seed are taken to constitute a sub-sample of approximately, but not less than, the required size.

Sample to be reduced Two spoons of different size

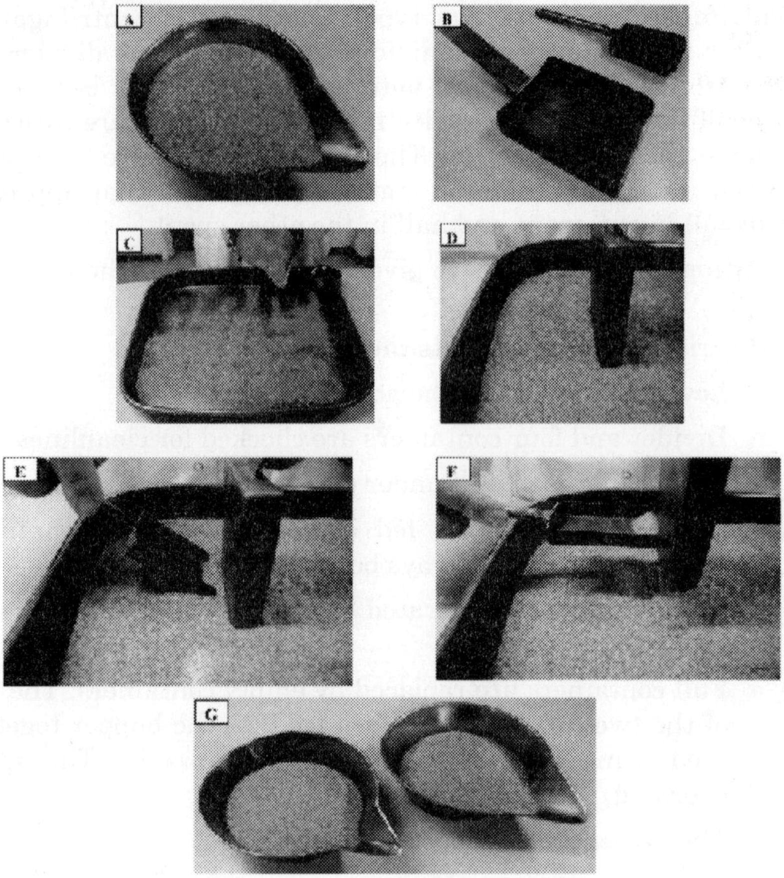

Distributing the seed over the pan. A spoon is pushed vertically into the seed layer (as a substitute for the spatula) E: With the second spoon the seed in front of the vertical spoon is collected. F: Both spoons are removed from the seed and the seed sample is transferred to a collection pan. G: Two sub-samples as the result.

(h) Hand halving method

In the International Rules this method is the most satisfactory method for chaffy and genera of tree and shrub seed:

Technique

1. The seed is **poured evenly** onto a smooth **clean** surface.
2. Thoroughly **mix the seed** into a mound with a flat-edged spatula
3. The mound is **divided into half** and each half is **halved again**, giving four portions. Each of the four portions is **halved again** giving eight portions which should be arranged in two rows of four.

4. **Combine and retain alternate portions**: e.g. combine the first and third portions in the first row with the second and fourth in the second row. Remove the remaining four portions.

5. **Steps two, three and four are repeated** using the retained portions from step four until the **weight** of sample **required** is obtained.

Sampling Mount Half with ruler

two halves four fourths final

Random Cups Method

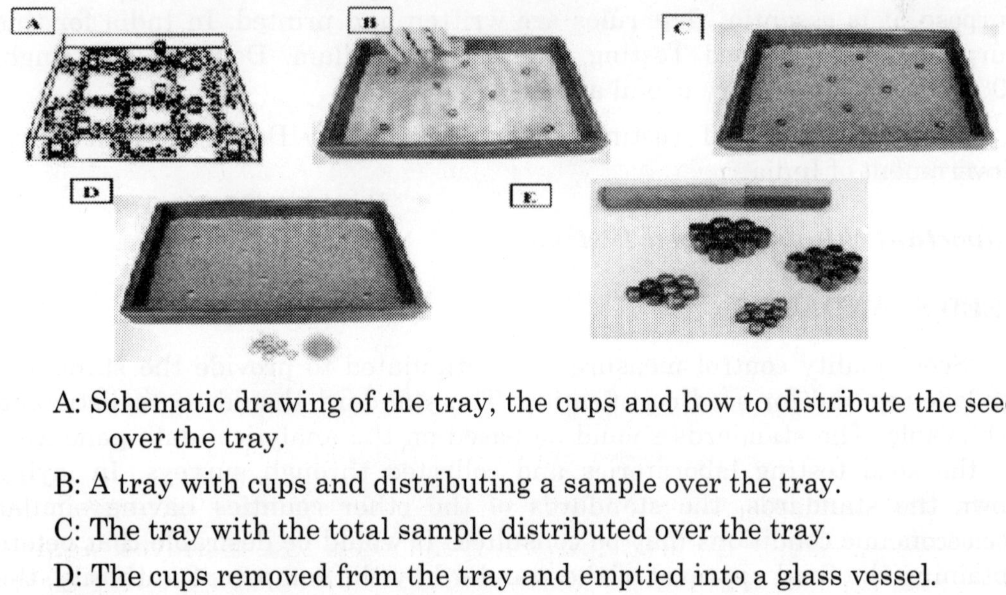

A: Schematic drawing of the tray, the cups and how to distribute the seed over the tray.

B: A tray with cups and distributing a sample over the tray.

C: The tray with the total sample distributed over the tray.

D: The cups removed from the tray and emptied into a glass vessel.

E: Cups of different size in one set.

Chapter - 11

Seed Testing

Importance of Seed testing, Seed Testing Laboratory and their role in seed certification.

INTRODUCTION

In large scale seed certification, involves the following three steps:-

1. Inspection of Seed Crop during its growth and Inspection of seed at processing plant.
2. Analysis of harvested and processed seed for quality attributes in Seed testing Laboratory.
3. Grow out Test to ascertain genetic purity.

Seed quality evaluation is done in the Seed Testing laboratories and they play a portal role in Seed Certification. In order to have uniformity in Seed testing it is important that the methods employed for testing, for this purpose, it is essential that rules are written and printed. In India for this purpose follow. A Seed Testing Manual by Challam, Douglas and Singh, 1967. Which served very useful and also followed.

Hand Book of Seed Testing, 1983. Published by Deptt. of Agriculture, Government of India.

Important things for Seed Testing

SEED STANDARDS

Seed quality control measures also stipulated to provide the standards for labeling and for seed certification. The standard should be realistic and achievable. The standards should be based on the analytical data generated by the seed testing laboratories and collected through surveys. In laying down the standards, the standards of the other counties having similar socioeconomic conditions may be consulted. It would be desirable that before obtaining the final approval of the standard by the competent authority, the standards should be discussed thoroughly by a committee, which is represented by the seed producers, certification and seed law enforcement officials. It would be desirable to prescribe two sets of standards namely, standards for labeling and standards for seed certification.

Seed certification standards may refer to the field condition or may refer to the procedures in inspection of the harvested produce as well as the manner of harvesting, transporting. Processing and packing. Unless, a seed certification agency keeps track of harvested produce until it is packed and sealed the identification of the lots cannot be assured. It is, therefore, necessary that the seed certification agency should lay down standard for processing plants. In addition, field and seed standards, such as isolation distances, inseparable other crop seeds, weeds, plants affected by seed borne diseases, genetic purity, percentage of pure seed, other crop and weed seeds, inert matter, moisture content, germination and insect damage should be prescribed for successful accomplishment of the certification.

For the purpose of labeling the seed under the seed law enforcement programme, the standard for pure seed and germination is usually required. These standards will vary accordingly to the kind of seed. The standards should be reviewed at timely intervals in order to make them realistic.

MEASUREMENT OF SEED QUALITY

It is absolutely important that the estimation of the quality of the seed lot should be made accurately. Wrong evaluation in seed quality measurement may cause problems in labeling, marketing and quality control operations. This may amount to large-scale rejection of the quality seed lots or approving sub-standard seed lot for seed certification or seed low enforcement. This will cause irreparable damage to the seed programme of the country. Accordingly, it is essential that the measurement of the seed quality should be made in accordance with prescribed procedure.

It is not possible to measure the genetic purity status of the seed lot accurately in a seed-testing laboratory. The official method for cultivars purity (Genetic Purity) is the "Grow-out-Test". The important laboratory methods for genetic purity testing includes the examination of the seed and seedlings, fluorescence test, phenol test, NaOH test, cotyledon color test and electrophoresis. These tests serve as complimentary methods for genetic purity determination. Grow-out-Test can determine the characteristics of the variety at field leavel.

The physical and physiological seed quality viz. physical purity, moisture content, germination and vigour is usually measured by employing the standardized procedures provided in the Rules for Seed Testing.

SAMPLING IN SEED TESTING LABORATORY

The seed samples, received in the laboratory are required to be reduced to obtain working samples as per crop for carrying out various tests. Seed sampling is aimed at obtaining a sample of the required size and consisting of the same components as the whole lot of processed seed. The quantity of Seed Sample for the laboratory is small, compared with the size of the seed lot, which it is intended to represent.

RECEIPT AND REGISTRATION OF SAMPLES

The receipt and registration of seed sample in a seed-testing laboratory is an important initial step for quality analysis of seed lots. This procedures help in identifying submitted samples and in carrying of the tests by of Certification procedural error to the maximum desired level.

There three types of seed samples received by a seed-testing laboratory. These are:

1 Service

2 Certification

3 Official Samples.

Separate "Entry books" may be maintained for the three categories of samples received. Different laboratory "test no. Series" may be used for each category. The samples shall be left in the incoming containers until they are entered and followed methods described in "Hand book of seed testing."

PHYSICAL PURITY ANALYSIS

The Physical Purity analysis seed sample refers to the determination of the different components of the physical purity viz. pure seed, other crop seeds, weed seeds and the inert matter.

The object of this test is to determine whether the submitted seed sample confirms to the prescribed quality standards in regard to purity components. Methods followed described in "Hand Book of the seed testing", 1993.

GERMINATION TESTING

Germination testing is the most important quality test in evaluating the planting value of a seed lot. The ability of seed to produce normal seedlings and plants later on is measured in terms of germination test in the laboratory hare been standardized to enable the test results to be reproduced within limits as nearly as possible as these determined by random sample variations. The objected for Germination test is to obtain information with respect to the planting value of the seed and to provide results which could be used to compare the value of different seed test.

Germination tests shall be made with seed from the pure seed fraction of a purity test. Minimums of four hundred seeds are required in four replicates of 100 each.

The seeds, arranged in require media in replicates, are lasted under favorable moisture conditions and in accordance with the methods prescribed in "Hand Book of Seed Testing", 1993.

MOISTURE DETERMINATION

Seed Moisture content is one of the important factor influencing seed quality and storability. So that, its estimation during seed quality determination is important. Seed moisture content can be expressed either on wet weight basis or dry weight basis. In seed testing, it is always expressed on a wet weight basis.

Seed Moisture content can be determined either by hot air oven or moisture meter. The methods for its calculation followed described in the Hand Book of Seed Testing, 1993.

STORING OF SAMPLES IN THE SEED TESTING LABORATORY

(a) Storage prior testing is undertaken storage, if necessary should be in a dry cool room.

(b) Storage after testing has been completed. The official samples after testing should be stored in controlled storage (5°C and 50% RH) for a minimum period of two years from the dated grant.

Table 11.1

S.N.	Crop	Maximum lot size (Qtls)	Minimum Submitted Sample (gms.)	Working Sample Size (gms.)	
				Purity analysis	Count of other Species
1.	**Cereals and Millets**				
	Barley	200	1000	120	1000
	Common millet	100	150	15	150
	Finger Millet	100	60	6	60
	Italian Millet	100	90	9	90
	Little millet	100	70	7	70
	Maize	400	1000	900	1000
	Oats	200	1000	120	1000

	Paddy	200	400	40	400
	Pearl Millet (Bajra)	100	150	15	150
	Rye	200	1000	120	1000
	Sorghum	100	900	90	900
	Wheat	200	1000	120	1000
2.	**Pulses**				
	Black gram	200	1000	700	1000
	Cowpea	200	1000	400	1000
	Gram	200	1000	1000	1000
	Green gram	200	1000	120	1000
	Khesari	200	1000	450	1000
	Lentil	100	600	60	600
	Red gram	200	1000	300	1000
3.	**Oilseeds**				
	Black Mustard	100	160	16	160
	Linseed	100	300	30	300
	Soyabean	200	1000	500	1000
	Castor	200	1000	1000	1000
	Groundnut, Indian mustard	200	1000	1000	1000
	Rapeseed	100	160	16	160
	Safflower	100	1000	180	1000
	Sesamum	100	70	7	70
	Sunflower	200	1000	250	1000
4.	**Cotton (hybrids)**	10	500	350	500
	Cotton (Varieties)	200	1000	350	1000
	Jute	100	150	15	150
	Mesta	100	700	70	700
	Sun hemp	100	700	70	700
5.	**Forages**				
	Berseem	100	60	6	60
	Lucerne	100	50	5	50
	Sudan grass	100	250	25	250
	Teosinte	200	1000	900	1000

6.	**Vegetables**				
(a)	Amaranthus	100	70	7	70
	Chinese cabbage	100	40	4	40
	Fenugreek (Methi)	100	40	4	40
	Lettuce	100	30	3	30
		100	500	50	500
(b)	Ash gourd	200	700	70	700
	Bitter gourd	200	1000	450	1000
	Bottle gourd	200	700	70	700
	Cucumber	100	150	70	150
	Long melon	100	150	70	150
	Muskmelon	100	150	70	150
	Pumpkin	100	350	180	350
	Ridge gourd	200	1000	400	1000
	Round gourd	200	1000	250	1000
	Sponge gourd	200	1000	350	1000
	Watermelon	200	1000	250	1000
	Summer squash	200	1000	700	1000
	Winter squash	200	1000	700	1000
(c)	Beet (all beets)	200	500	50	500
	Carrot	100	30	3	30
	Radish	100	300	30	300
	Turnip	100	70	7	70
(d)	Bhindi	200	1000	140	1000
	Brinjal, capsicum, chilies	100	150	15	150
	Tomato	100	70	7	70
(e)	Brussels sprouts, cabbage, Cauliflower, broccoli karamsag, Knol-Knol, Kohl rabi, Sprouting broccoli	100	100	10	100
(f)	Cluster bean	200	800	80	800
	Dolichos bean	200	1000	600	1000
	French bean	200	1000	1000	1000
	Garden bean	200	1000	1000	1000
(g)	Onion	100	80	80	80

Table 11.2 General seed standards prescribed as per MSCS

Rabi crops

Crop	Stage	Purity	Other Crop Seed	Weed Seed	ODV Other Dist. Variety	Obj. Weed Seeds	Inert Mater %	Moisture %	Germination %
Wheat	F/S	98	10/kg	10/kg	-	2/kg	2	12	85
	C/S	98	20/kg	20/kg	-	5/kg	2	12	85
Barley	F/S	98	10/kg	10/kg	10/kg	2/kg	2	12	85
	C/S	98	20/kg	20/kg	20/kg	5/kg	2	12	85
Gram	F/S	98	0 /kg	0	5/kg	-	2	9	85
	C/S	98	5 /kg	0	10/kg	-	2	9	85
Mustard	F/S	97	10/kg	10/kg	0.1%	5/kg	2	12	85
	C/S	97	20/kg	20/kg	0.50%	10/kg	2	12	85
Maize	F/S	98	5/kg	0	10/kg	-	2	12	90
	C/S	98	10/kg	0	20/kg	-	2	12	90
Oat	F/S	98	10/kg	10/kg	10/kg	2/kg	2	12	85
	C/S	98	20/kg	20/kg	20/kg	5/kg	2	12	85
Berseem	F/S	98	10/kg	10/kg	-	5/kg	2	10	80
	C/S	98	20/kg	20/kg	-	10/kg	2	10	80
Palak	F/S	96	5/kg	5/kg	-	-	4	9	60
	C/S	96	10kg	10/kg	-	-	4	9	60

Table 11.3 Seed Standards of the general Crops

Kharif crops

Crop	Stage	Purity	Other Crop Seed	Other distinguishable Variety	Weed Seed	objectionable Seed	Inert Matter	Moisture	Germination %
Bajra	F/S	98	10	-	10	-	2	12	75
	C/S	98	10	-	20	-	2	12	75
Jowar	F/S	98	5	10	5	-	2	12	75
	C/S	98	10	20	10	-	2	12	75
Maize	F/S	98	5	5	0	-	2	12	75
	C/S	98	10	10	0	-	2	12	80-90%
Paddy	F/S	98	10	10	10	2	2	13	80
	C/S	98	20	20	20	5	2	13	80
Cotton	F/S	98	5	-	5	-	2	10	65
	C/S	98	10	-	10	-	2	10	65
Guar	F/S	98	10	10	51	-	2	12	75
	C/S	98	20	20	10	-	2	12	75
Arhar	F/S	98	5	10	5	-	2	9	75
	C/S	98	10	20	10	-	2	9	75

Moong	F/S	98	5	10	5	-	2	9	75
	C/S	98	10	20	10	-	2	9	75
Moth	F/S	98	5	10	5	-	2	9	75
	C/S	98	10	20	10	-	2	9	75
Urd	F/S	98	5	10	5	-	2	9	75
	C/S	98	10	20	10	-	2	9	75
Cowpea	F/S	98	5	5	10	-	2	9	75
	C/S	98	10	10	10	-	2	9	75
G.Nut	F/S	96	5	-	0	-	4	9	70
	C/S	96	10	-	0	-	4	9	70
Soybean	F/S	98	5	0	10	-	2	12	70
	C/S	98	10	10	40	-	2	12	70
Castor	F/S	98	0	5	0	-	2	8	70
	C/S	98	0	10	0	-	2	8	70
Til	F/S	98	10	10	10	-	3	9	80
	C/S	97	20	20	20	-	3	9	80

SEED HEALTH TESTS

Many high yielding priorities have shown susceptibility to different diseases and many of these diseases are seed borne. A pathogen may penetrate several of these parts of the seed and in turn infect them. The infestation/ contamination of the seed may occur during harvesting, thrusting and processing.

The seed borne pathogens may result in loss in germination, discoloration and shriveling, development of plant diseases, distribution of pathogen to new areas, introduction of new strains, toxin production in infected seed etc. For the testing of diseases methods have been described in handbook of seed testing, 1993.

DETERMINATION OF INSECTS ETC.

To determine insects, nematodes and plant parasites in the seed sample or planting material and by inference that the seed lot. The information thus obtained will be useful in detection of infestation elucidates the cause of poor germination, weak seedling etc., Hidden infestation leads for further infestation of seed lot during transit or storage under ambient condition and it helps in adoption of proper remedial measures.

Chapter - 12

Seed Marketing

INTRODUCTION

Seed marketing is an another important factor in seed industry. Each company has its own strategy of selling the products. The following points are to be considered for best marketability of the products.

1. To ensure the best quality of seed in the market, which can compete the other products.
2. To form Wholeseller /Retailers in the Market
3. The policy to give commissions to Wholeseller/Retailers in the Market must be in such a manner so that they cannot deviate from your company to other.
4. Suitable feedback is to be provided to them so as the actual information about the product is there in the market.
5. Time to time some seminars are to be conducted with wholeseller / Retailers to rectify their problems if any.
6. Time to time some Field Day's are to be conducted Feedback of the farmers are to be taken the wholeseller/Retailers must also be involved.
7. The proper advertisements must be done in the market about the products.
8. Own research must be developed to stay in the market as monopoly items.
9. To attend complaint immideately if any regarding farmer/dealer/ retailer.
10. Field complaints action must be taken immidately regarding germination.

Chapter 13

Seed Legislations

The SEEDS ACT, 1966
(Act No. 54 of 1966)

An Act to provide for regulating the quality of certain seeds for sale, and for matters connected therewith.

Be it enacted by Parliament in the Seventeenth Year of the Republic of India as follows:-

Short Title, extent and commencement

1. (1) This Act may be called the Seeds Act, 1966

(2) It extends to the whole of India.

(3) It shall come into force on such date as the Central Government may, by notification in the official Gazette, appoint, and different dates may be appointed for different provisions of this Act, and for different States or for different areas thereof.

Definitions

2. In this Act, unless the context requires:-

1. "Agriculture" includes horticulture.
2. "Central Seed Laboratory" means the Central Seed Laboratory established or declared as such under sub section (1) of section 4.
3. "Certification agency" means the certification agency established under section 8 or recognized under section 18.
4. "Committee" means the central seed committee constituted under sub section (1) of section 31.
5. "Container" means a box, bottle, casket, tin, barrel, case, receptacle, sack, bag, wrapper or other thing in which any article or thing is placed or packed.
6. "Export" means taking out of India to a place outside India.
7. "Import" means bringing into India from a place outside India.

8. "Kind" means one or more related species or sub-species of crop plants each individually or collectively known by one common name such as cabbage, maize, paddy, and wheat.
9. "Notified kind or variety" in relation to any seed, means nay kind or variety thereof notified under section 5.
10. "Prescribed" means prescribed by rules made under this Act.
11. "Seed" means any of the following classes of seeds used for sowing or planting-
    a. Seed of food crops including edible oil seeds and seeds of fruits and vegetables.
    b. Cotton Seeds.
    c. Seed of cattle fodder.
    d. And includes seeding and rubbers, bulbs, rhizomes, roots, cuttings, all types of grafts and other vegetatively propagated material, of food crops of cattle fodder.
12. "Seed Analyst" means a seed analyst appointed under section 12.
13. "Seed Inspector" means a Seed Inspector appointed under section 13.
14. "State Government" in relation to a Union territory, means the administrator thereof.
15. "State Seed Laboratory" in relation to any state, means the state seed laboratory established of declared as such under sub section (2) of section 4 for that state and
16. "Variety" means a sub-division of a kind identifiable by growth, yield, plant fruit, seed or other characteristic.

Central Seed Committee

3.

1. The Central Government shall, as soon as may be after the commencement of this act, constitute called the central seed committee to advise the central Government and the state governments on matters arising out of the administration of this act and to carry out the other functions assigned to it by or under this Act.
2. The committee shall consist of the following members, namely:-
    a. A chairman to be nominated by the Central Government.
    b. Eight persons to be nominated by the central Government to represent such interests as that Government thinks fit, of whom not less than two persons shall be representatives of growers of Seed:-

3. One person to be nominated by the Government of each of the States.
4. The members of Committee shall unless their seats become vacant earlier by resignation, death or otherwise, be entitled to hold office for two years and shall be eligible for re-nomination.
5. The Committee may, subject to the previous approval of the central government, make byelaws fixing the quorum and regulating its own procedure and the conduct of all business to be transacted by it.
6. The Committee mat appoint one or more sub-committees, consisting wholly of members of the Committee or wholly of other persons or partly of members of the Committee and partly of other persons, as it thinks fit, for the purpose of discharging such of its functions as may be delegated to such sub committee or sub committees by the Committee.
7. The functions of the Committee or any sub-committee thereof mat be exercised not-withstanding any vacancy therein.
8. The Central Government shall appoint a person to be the secretary of the Committee and shall provide the Committee with such clerical and other staff as the Central Government considers necessary.

4.

1. The Central Government may be notification in the official Gazette, establish central seed laboratory of declare any seed laboratory as the Central seed Laboratory to carry out the functions entrusted to the central Seed Laboratory by or under this Act.
2. The State Government may, be notification in the Official Gazette, establish one or more state Seed Laboratories or declare any seed laboratory as a state seed laboratory where analysis of seeds of any notified kind or variety shall be carried out by seed analysts under this Act in the prescribed manner.

Power to notify kinds or varieties of seeds

5. If the Central Government, after consultation with the Committee, is of opinion that it is necessary or expedient to regulate the quality of seed of any kind or variety to be sold for purposes of agriculture, it may, be notification in the Official Gazette, declare such kind or variety to be a notified kind or variety for the purposes of this Act and different kinds or varieties may be notified for different states or for different area thereof.

Power to specify minimum limits of germination and purity, etc.

6. The Central Government of germination and purity with respect to ant seed of any notified kind or variety;

(a) The minimum limits of germination and purity with respect to any seed of any notified kind or variety.

(b) The mark or label to indicate that such seed conforms to the minimum limits of germination and purity specified under clause (a) and the particulars which such mark or label may contain.

Regulation of sale of seeds of notified kinds or varieties

7. No person shall, himself or by any other person on his behalf, carry on the business of selling, keeping for sale, offering to sell, bartering or otherwise supplying any notified kind or variety, unless:-

(a) Such seed is identifiable as tot its kind or variety;

(b) Such seed conforms to the minimum limits of germination and purity specified under clause (a) of section 6.

(c) The container of such bears in the prescribed manner, the make or label containing the correct particulars thereof, specified under clause (b) of section 6; and

(d) He complies with such other requirements as may be prescribed.

Certification agency

8. The State Government of the Central Government in consultation with the State Government may, by notification in the Official Gazette, establish a certification agency for the State to carry out the functions entrusted to the certification agency by or under this Act.

Grant of certificate by certification agency

9.

1. Any person selling, keeping for sale, offering to sell, bartering or otherwise supplying any seed of any notified kind or variety may, if he desires to have such seed certified by the certification agency, apply to the certification agency for the grant of a certificate for the purpose.

2. Every application under sub-section (1) shall be made in such form, shall contain such particulars and shall be accompanied by such fees as may be prescribed.

3. On receipt of ant such application for the grant of a certificate, the certification agency may, after such enquiry as it thinks fit and after satisfying itself that the seed to which the application relates conforms to the minimum limits of germination and purity specified for that seed under clause (a) of section 6, grant a

certificate in such form and on such conditions as may be prescribed.

Revocation of certificate

10. If the certification agency is satisfied, either on a reference made to it in this behalf or otherwise that:-

(a) The certificate granted by it under section 9 has been obtained by misrepresentation as to an essential fact; or

(b) The holder of the certificate has, without reasonable cause, failed to comply with the conditions subject to which the certificate has been granted or has contravened any of the provisions of this Act or the rules made there under;

Then, without prejudice to any other penalty to which the holder of the certificate may be liable under this Act, the certification agency may, after giving the holder of the certificate an opportunity of showing cause, revoke the certificate.

Appeal

11.

1. Any person aggrieved by a decision of a certification agency under section 9 or section 10 may, within thirty days from the date on which the decision is communicated to him and on payment of such fees as may be prescribed, prefer an appeal to such authority as may be specified by the State Government in this behalf;
2. Provided that the appellate authority may entertain an appeal after the expiry of the said period of thirty days if it is satisfied that the appellant was prevented by sufficient cause from filling the appeal in time.
3. On receipt of an appeal under sub-section (1), the appellate authority shall, after giving the appellant an opportunity of being heard, dispose of the appeal as expeditiously as possible.
4. Every order of the appellate authority under this section shall be final.

Seed Analysts

12. The State government may, by notification in the Official Gazette, appoint such persons as it thinks fit, having the prescribed qualifications, to be Seed Analysts and define the areas within which they shall exercise jurisdiction.

Seed Inspectors

13.

1. The State Government may, by notification in the official Gazette, appoint such persons as it thinks fit, having the prescribed qualifications, to be Seed inspectors and define the areas within which they shall jurisdiction.
2. Every Seed Inspector shall be deemed to be a public servant within the meaning of section 21 of the Indian Penal Code (45 of 1860) and shall be officially subordinate to such authority as the State Government may specify in this behalf.

Powers of Seed Inspector

14.

1. The Seed Inspector m.
   a. Take samples of any seed of any notified kind or variety from-
      i. Any person selling such seed or
      ii. And person who is in the course of covering, delivering or preparing of deliver such seed to a purchaser or a consignee; or
      iii. A purchaser or a consignee after delivery of such seed to him;
   b. Send such sample for analysis to the Seed analyst for the area within which such sample has been taken;
   c. Enter and search at all reasonable times, with such assistance, if any, as he considers necessary, any place in which he has reason to believe that an offence under this Act has been or is being committed and order in writing the person in possession of any seed in respect of which the offence has been or is being committed, not to dispose of any stock of such seed for a specific period not exceeding thirty days or, unless the alleged offence is such that the defect may be removed by the possessor of the seed seize the stock of such seed;
   d. Examine any record, document or any other material object found in any place mentioned in clause (c) and seize the same if he has reason to believe that it may furnish evidence of the commission of an offence punishable under this Act and

e. Exercise such other powers as may be necessary for carrying out the purposes of this Act or any rule made there under.

2. Where any sample of any seed of any notified kind or variety is taken under clause (a) of sub section (1), its cost, calculated at the rate at which such seed is usually sold to the public, shall be paid on demand to the person from whom it is taken.

3. The power conferred by this section includes power to break open any container in which any seed of any notified kind or variety may be contained or to break open the door of any premises where any such may be kept for sale;

4. Provided, that the power to break open the door shall be exercised only after the owner or any other person in occupation of the premises, if he is present therein, refuses to open the door on being called upon to do so.

5. Where to seed Inspector takes any action under clause (a) of sub section (1), he shall, as far as possible, call not less than two persons to be present to be present at the time when such action is taken and take their signatures on a memorandum to be prepared in the prescribed form and manner.

6. The provisions of the code of criminal procedure, 1898 (5 of 1898) shall, so far as may be, apply to any search orb seizure under this section as they apply to any search or seizure made under the authority of a warrant issued under section 98 of the said Code.

Procedure to be followed by Seed Inspectors

15.

1. Whenever a Seed Inspector intends to take sample of any seed of any notified kind or variety for analysis, he shall-
   a. Given notice in writing, then an there, of such intention to the person from whom he intends to take sample,
   b. Except in special cases provided by rules made under this Act, take three representative samples in the prescribed manner and mark and seal of fasten up each sample in such manner as its nature permits.

2. When samples of any seed of any notified kind or variety are taken under sub section (1), the Seed Inspector shall-
   a. Deliver one sample to the person from whom it has been taken;

b. Send in the prescribed manner another sample for analysis to the seed Analyst for the area within which such sample has been taken, and

c. Retain the remaining sample in the prescribed manner for production in case any legal proceedings are taken or for analysis by the Central Seed Laboratory under sub section (2) of section 16, as the case mat be.

3. If the person from whom the samples has been refuses to accept one of the samples, the Seed Inspector shall send intimation to the Seed Analyst of such refusal and thereupon the Seed Analyst reserving the sample for analysis shall divide it into two parts and shall seal or fasten up one of those parts and shall cause it, either upon receipt of the sample or when he delivers his report, to be delivered to the Seed Inspector who shall retain it for production in case legal proceedings are taken-

4. Where a Seed Inspector takes any action under clause (c) of sub section (1) of section 14;

a. He shall use all dispatch in ascertaining whether or not the seed contravenes any of the provisions of section 7 and if it is ascertained that the seed does not so contravene, forthwith revoke the order passed under the said clause or, as the case may be taken such action as may be necessary for the return of the stock of the seed seized.

b. If he seized the stock of the seed, he shall, as soon as may be, inform a magistrate and take his orders as to custody thereof;

c. Without prejudice to the institution of any prosecution, if the alleged offence is such that the defect may be removed by the possessor of the seed, he shall, on being satisfied that the defect has been so removed, forthwith revoke the order passed under the said clause.

5 Where a Seed Inspector seized any record, register, document or any other material object under clause (d) of sub section (1) of section 14, he shall, as soon as may be, inform a magistrate and take his orders as to the custody thereof.

Report of Seed Analyst

16.

1. The Seed Analyst shall, as soon as may be after the receipt of the sample under sub section (2) of section 15, analyses the sample at the State Seed Laboratory and deliver, in such form as may be

prescribed, one copy of the report of result of the analysis to the Seed Inspector and another copy thereof to the person from whom the sample has been taken.

2. After the institution of a prosecution under this Act, the accused vendor or the complainant may, on payment of the prescribed fee, make an a application to the court for sending any of the samples mentioned in clause (a) or clause (c) of sub section (2) of section 15 to the Central Seed Laboratory for its report and on receipt of the application, the court shall first ascertain that the mark and the seal or fastening as provided in clause (b) of sub section (1) of section 15 are intact and may then dispatch the sample under its own seal to the Central Seed Laboratory which shall thereupon send its report tot the court in the prescribed form within on month from the date of receipt of the sample, specifying the result of the analysis.

3. The report sent by the Central Seed Laboratory under sub section (2) shall supersedes the report given by the Seed Analyst under sub section (1).

4. Where the report sent by the Central Seed Laboratory under sub section (2) is produced in any proceedings under section 19, it shall not be necessary in such proceedings to produce any sample or part thereof taken for analysis.

Restriction on exports and import of seeds of notified kinds or varieties.

17. No person shall, for the purpose of sowing or planting by any person (including himself) export or import or cause to be exported or imported any seed of any notified kind or variety, unless:-

(a) It conforms to the minimum limits of germination and purity specified for that seed under clause (a) of section 6; and

(b) Its container bears, in the prescribed manner, the mark or label with the correct particulars there of specified for that seed under clause (b) of section 6.

Recognition of Seed Certification agencies of foreign countries

18. The Central Government may, on the recommendation of the Committee and by notification in the Official Gazette, recognize any seed certification agency established in any foreign country, for the purposes of this Act.

Penalty

19. If any person:-

- **(a)** Contravenes any provision of this Act or any rule made there under; or
- **(b)** Prevents a seed Inspector from taking sample under this Act, or
- **(c)** Prevents a Seed Inspector from exercising any other power conferred on him by or under this Act.

He shall on conviction, be punishable-

- (i) For the first offence with fine which may extend to five hundred rupees, and
- (ii) In the event of such person having been previously convicted of an offence under this section, with imprisonment for a term which may extend to six months, or with fine which mat extend to one thousand rupees, or with both.

Forfeiture of property

20. When any person has been convicted under this Act for the contravention of any of the provisions of this Act or the rules made there under, the seed in respect of which the contravention has been committed may be forfeited to the Government.

Offences by companies

21.

- **(a)** Where an offence under this act has been committed by a company, every person who at the time the offence was committed was in charge of, and was responsible to the company for the conduct of the business of the company, as well as the company, shall be demand to be guilty of the offence and shall be liable to be proceeded against and punished accordingly.

Provided that nothing contained in this sub-section shall render any such person liable to any punishment under this Act if he proves that the offence was committed without his knowledge and that he exercised all due diligence to prevent the commission of such offence.

- **(b)** Notwithstanding anything contained in sub section (1), where an offence under this Act has been committed by a company and it is proved that the offence has been committed with the consent or connivance of, or is attributable to any neglect on the part of, any director, manager, secretary or other of the company, such director, manager, secretary or other officer shall also be deemed to be guilty of that offence and shall be liable to be proceeded against and punished accordingly.

Explanation : For the purposes of this Section-

(a) 'Company' means any body corporate and includes a firm or other association of individuals; and

(b) 'Director', in relation to a firm, means a partner in the firm.

Protection of action taken in good faith

22. No suit, prosecution or other legal proceeding shall lie against the Government or any officer of the Government for anything which is in good faith done or intended to be done under this Act.

Power to give directions

23. The Central Government may given such directions to any State Government as may appear to the Central Government to be necessary for carrying into execution in the State any of the provisions of this Act or of any rule made there under.

Exemption

24. Nothing in this Act shall apply to any seed of any notified kind or variety grown by a person and sold or delivered by him on his own premises direct to another person for being used by that person for the purpose of sowing or planting.

Power to make rules

25.

1. The Central Government any, by notification in the Official Gazette, make rules to carry out the purposes of this Act.
2. In particular and without prejudice to the generality of the fore-going power, such rules may provide for-
   a. The functions of the Committee and the traveling and daily allowances payable to members of the Committee and members of any sub committee appointed under sub section (5) of section 3;
   b. The function of the Central Seed Laboratory;
   c. The functions of a certification agency;

d. The manner of marking or labeling the container of seed of any notified kind or variety under clause (c) of section 7 and under clause (b) of section 17.

e. The requirements, which may be complied with by a person carrying on, the business referred to in section 7.

f. The form of application for the grant of a certificate under section 9, the particulars it may contain, the fees which should accompany it, the form of the certificate and the conditions subject to which the certificate may be granted.

g. The form and manner in which and the fee on payment of which an appeal may be preferred under section 11 and the procedure to be followed by the appellate authority in disposing of the appeal;

h. The qualifications and duties of Seed Analysts and Seed Inspectors;

i. The manner in which samples may be taken by the Seed Inspector, the procedure for sending such samples to the Seed Analyst or the Central Seed Laboratory and the manner of analyzing such samples;

j. The form of report of the result of the analysis under sub section (1) or sub section (2) of section 16 and the fees payable in respect of such report under the said sub section (2).

k. The records to be maintained by a person carrying on the business referred to in section 7 and the particulars which such records shall contain; and

l. Any other matter which is to be or may be prescribed.

3. Every rule made under this Act shall be laid as soon as after it is made, before each House of Parliament while it is in session for a total period of thirty days which may be comprised in one session or in two successive sessions, and if, before the expiry of the session in which it is so laid or the session immediately following, both Houses agree in making any modification in the rule or both Houses agree that the rule should not be made, that rule shall, thereafter have effect only in such modified form or be of no effect, as the case may be; so. However, that any such modification or annulment shall be without prejudice to the validity of anything previously done under that rule.

THE SEED RULE

THE SEED RULES 1968

PART 1 –Preliminary

1. Short title: These rules may be called the Seed Rules, 1968.

2. Definitions: In these rules, unless the context otherwise require;-

(a) "Act" means the Seeds Act, 1966 (54 of 1966);

(b) "Advertisement" means all representations other than those on the label, disseminated in any manner or by any means relating to seed for purposes of the Act;

(c) "Certification Sample" means a sample of seed drawn by a certification agency or by a duly authorized representative, of a certification agency established under section 8 or recognized under section 18 of the Act.

(d) "Certification tag" means a tag or label of certain design to be specified by the certification agency and shall constitute the certificate granted by the certification agency.

(e) "Certified Seed" means a person who grows or distributes certified seed in accordance with the procedure and to the container of which the certification tag is attached;

(f) "Certified seed producer" means a person who grows or distributes certified seed in accordance with the procedure and standards of the certification agency;

(g) "Complete Record" means the information, which relates to the origin, variety, kind, germination and purity of seed of ant notified kind or variety offered for sale, sold or otherwise supplied.

(h) "Form" means a from appended to these rules:

(i) "Origin" means the State, Union Territory or foreign country where the seed is grown and in case seeds of different origin are blended the label shall show the percentage of seed of each origin.

(j) "Processing" means cleaning, drying, treating, grading and other operations which would change the purity and germination of the seed and thus requiring re testing to determine the quality of the seed, but does not include operations such as packaging and labeling;

(k) "Section" means a section of the Act;

(l) "Service Sample" means a sample to the Central Seed Laboratory or to a State Seed Laboratory for testing, the results to be used as information for seeding, selling or labeling purposes;

(m) "Treated" means that the seed has been subjected to an application of a substance or process in such a manner as to reduce, control or repel certain disease organisms, insects, or any other pests attacking such seed or seedlings growing there from and for other purposes.

PART II Central Seed Committee

3. Functions of the Central Seed Committee: In addition to the functions entrusted to the Committee by the Act, the Committee shall-

(a) Recommend the rate of fees to be levied for analysis of samples by the Central and State Seed Testing Laboratories and for certification by the Certification agencies;

(b) Advise the Central or State Governments on the suitability of seed testing laboratories;

(c) Send its recommendations and concerning records to the Central Government;

(d) Recommend the procedure and standards for certification, tests and analysis of seeds, and

(e) Carry out such other functions as are supplemental, incidental or consequential to any of the functions conferred by the Act or these rules.

4. Traveling and Daily Allowances Payable to Members of the Committee and its sub committees:- The members of the committee and its sub committees shall be entitled to draw traveling and daily allowances as specified below when they are called upon to attend a meeting of the committee or a sub-committee thereof:-

(a) An official member of the committee or its sub committee shall be entitled to draw traveling and daily allowances in accordance with the rules of the Government under which he is for the time being employed and from the same source from which his pay and allowances are drawn.

(b) A non-official member shall be allowed traveling and daily allowances in accordance with the general orders issued in this behalf by the Central Government from time to time.

PART III - Central Seed Laboratory

5. Functions:- In addition to the functions entrusted to the Central Seed Laboratory by the Act, the Laboratory shall carry out the following functions, namely:-

(a) Initiate testing programme in collaboration with the State Seed Laboratories designed to promote uniformity in test results between all seed laboratories in India.

(b) Collect data continually on the quality of seed found in the market and make this data available to the Committee; and

(c) Carry out such other functions as may be assigned to it by the Central Government from time to time.

PART IV - Seed Certificate Agency

6. Functions of the Certification Agency:- In addition to the functions entrusted to the certification agency by the Act, the Agency shall-

(a) Certify seed of any notified kinds or varieties;

(b) Outline the procedure for submission of applications such for growing, harvesting, processing, storage and labeling of seeds intended for certification till the end to ensure that seed lots finally approved for certification are true to variety and meet prescribed standards for certification under the Act or these rules;

(c) Maintain a list of recognized breeders of seeds;

(d) Verify, upon receipt of an application for certification that the variety is eligible for certification that the seed source used for planting was authenticated and the record of purchase is in accordance with these rules and the fees have been paid.

(e) Take sample and inspect seed lots produced under the procedure laid down by the certification agency and have such samples tested to ensure that the seed conforms to the prescribed standards of certification;

(f) Inspect seed processing plants to see that the admixtures of other kinds and varieties are not introduced.

(g) Ensure that action at all stages, e.g. field inspection, seed processing plant inspection, analysis of samples taken and issue of certificates (including tags, marks, labels and seals) is taken expeditiously.

(h) Carry out educational programme designed to promote the use of certified seed including a publication listing certified seed growers and sources of certified seed;

(i) Grant certifications (including tags labels, seals etc.) in accordance with the provisions of the Act and these rules;

(j) Maintain such records as may be necessary to verify that seed plans for the production of certified seed were eligible for such planting under these rules;

(k) Inspect fields to ensure that the minimum standards for isolation, rouging (where applicable) use of male sterility (when applicable) and similar factors are maintained at all times, as well as ensure that seed borne diseases are not present in the field to a granter extent than those provided in the standards for certification.

PART V - Marking or Labeling

7. Responsibility for Marking or Labeling : When seed of a notified kind or variety is offered for sale under section 7, each container shall be marked or labeled in the manner hereinafter specified. The person whose name appears on the mark or label shall be responsible for the accuracy of the information required to appear on the mark or label so long as seed is contained in the unopened original container.

Provided, however, that such person shall not be responsible for the accuracy of the statement appearing on the mark or label if the seed is removed from the original unopened container. Or he shall not be responsible for the accuracy of the germination statement beyond the date of validity indicated on the mark or label.

8. Contents of the mark or label : There shall be specified on every mark or label-

(i) Particulars, as specified by the Central Government under clause (b) of section 6 of the Act;

(ii) A correct statement of the net content in terms of weight and expressed in metric system;

(iii) Date of testing;

(iv) If the seed in container has been treated-

a. A statement indicating that the seed has been treated;

b. The commonly accepted chemical or abbreviated chemical (generic) name of the applied substance; and

c. If the substance of the chemical used for treatment, and present with the seed is harmful to human beings or other vertebrate animals, a caution statement such as "Do not use for food, feed or oil purposes". The caution for mercurial and similarly toxic substance shall be the word "Poison" which shall be in type size, prominently displayed on the label in red;

(v) The name and address of the person who offers for sale, sells or otherwise supplies the seed and who is responsible for its quality;

(vi) The name f the seed as notified under section 5 of the Act.

9. Manner of Marking or labeling the container under clause (C) of section 7 and clause (B) of Section 17 :

1. The mark or label containing the particulars of the seed as a specified under clause (b) of section 6 shall appear on each container of seed or on a tag or mark or label attached to the container in a conspicuous place on the innermost container in which the seed is packed and on every other covering in which that container is packed and shall be legible.
2. Any transparent cover or any wrapper, case or other covering used solely for the purpose of packing of transport or delivery need not be marked or labeled.
3. Where by a provision of these rules, any particulars are required to be displayed on a label on the container, such particulars may, instead of being displayed on a label be etched painted or otherwise indelibly marked on the container.

10. Mark or label not to contain false or misleading statement:- The mark or label shall not contain any statement, claim design, device, fancy name or abbreviation which is false or misleading in any particular concerning in the seed contained in the container.

11. Mark or label not to contain reference to the Act or Rules contradictory to required particulars:- The mark or label shall not contain any reference to the Act, or any of these, rules or any comment on, or reference to, or explanation of any particulars or declaration required by the Act or any of these rules which directly or by implication contradicts, qualifies on modifies such particulars or declaration.

12. Denial of Responsibility for mark or label content prohibited:- Nothing shall appear on the mark or label or in any advertisement pertaining to any seed of any notified kind or variety which shall deny responsibility for the statement required by or under the Act to appear on such mark, label or advertisement.

PART VI- Requirements

13. Requirements to be complied with by a person carrying on the Business referred to in Section 7.

1. No person shall sell, keep for sale, offer to sell, barter or otherwise supply any seed of any notified kind or variety, after the date recorded on the container, mark or label as the date up to which the seed may be expected to retain the germination not less than that prescribed under clause (a) of section 6 of the Act.
2. No person shall alter, obliterate or deface any mark or label attached to the container of any seed.
3. Every person selling, keeping for sale, offering to sell, bartering or otherwise supplying any seed of notified kind or variety under section 7, shall keep over a period of three years a complete record of each lot of seed sold except that any seed sample may be discarded one year after the entire lot represented by such sample has been disposed of. The sample of seed kept as part of the complete record shall be as large as the size notified in the official Gazette. This sample, if required to be tested, shall be tested only for determining the purity.

14. Classes and source of Certified Seed :

1. There shall be three classes of certified seed, namely, foundation, registered and certified and each class meet the following standards for the class:-
   a. Foundation seed shall be the progeny of breeders' seed, or be produced from foundation seed, which can be clearly traced to breeder's seed. Production shall be supervised and approved by a seed certification agency and be so handled as to maintain specific genetic purity and identity and shall be required to meet certification standards for the crop being certified.
   b. Registered seed shall be the progeny of registered or foundation seed that is so handled as to maintain its genetic identity and purity according to standard specified for the particular crop being certified.
   c. Certified seed shall be the progeny of registered or foundation seed that is so handled to maintain genetic identity and purity according to standards specified for the particular crop being certified.
2. At the discretion of the certification agency (when considered necessary to maintain adequate seed supplies) certified seed may be the progeny of certified seed provided this reproduction may not exceed three generations and provided further that it is determined by the seed certification agency, that the genetic purity will not be significantly altered.

PART VII- Certification of Seeds

15. Application for grant of a Certificate : Every application for the grant of a certificate under sub-section (1) of section 9 shall be made in Form 1 in accordance with the procedure outlined by the certification agency for submission of applications and contain the following particulars, namely :

(a) The name, profession, and place of residence of the applicant;

(b) The name, of the seed to be certified, its notified kind or variety;

(c) Class of the Seed;

(d) Source of the Seed;

(e) Limits of germination and purity of the seed;

(f) Mark or label of the seed.

16. Fees: Every application under sub-section (1) of section 9 shall be accompanied by a fee of Rs. 25 in Cash.

17. Certificate: Every certificate granted under sub section (3) of section 9 shall be in From 11 and shall be granted by the certification agency, after making enquiries and satisfying itself in accordance with the provisions of the said sub section on the following conditions, for the period to be specified by the certification agency, namely :

(I) The person to whom the certificate is granted under sub section (3) of section 9 shall attach a certification tag to every container of the certified seed and shall follow the provisions in respect of marking or labeling provided by or under the Act.

(II) The Certification tag shall contain the following particulars, namely;

a. Name and address of the certification agency;

b. Kind and variety of the seed;

c. Lot number or other mark of the seed;

d. Name and address of the certified seed producer;

e. Date of issue of the certified and of its validity;

f. An appropriate sign to designate certified seed;

g. An appropriate word denoting the class designation of the seed.

(III) The Colour of the certification tag shall be white for foundation seed, purple for registered seed and blue for certified seed.

(IV) The container of the certified seed shall carry a seal of such material and in such form as the certification agency may determine and no container carrying a certification tag shall be sold by the person if the tag or seal has either been tampered with or removed.

(V) The Certification tag on the container shall specify;

a. The period during which the seed shall be used for sowing or planting;

b. That the use of seed after the expiry of the validity period by any person is entirely at his risk and the holder of the certificate shall not be responsible for any damage to the buyer of the seed;

c. That no one should purchase the seed if the seal or the certification tag has been tampered with.

(VI) The holder of the certificate shall keep record of the details of each lot of the seed which is issued for sale in such form as to be available for inspection and to be easily identified by reference to the number of the lot as shown in the certification tag of each container and such records shall be retained in the case of a seed for which expiry date is fixed for a period of two years from the expiry of such date.

(VII) The holder of the certificate shall allow any Seed Inspector, authorized in writing by the certification agency in that behalf, to enter with or without prior notice the premises where the seeds are grown processed and sold and to inspect premises, plant and the process of processing at all reasonable hours.

(VIII) The holder of the certificate shall allow the Seed Inspector, authorized in writing by the certification agency, to inspect all registers and records maintained under these rules and to take samples of the seeds and shall supply to the Seed Inspector such information as he may require for the purpose of ascertaining whether the conditions subjects to which the certificate has been granted, have been complied with.

(IX) The holder of the certificate shall on request furnish to the certification agency from every lot of the seed or from such lot or lots as the said agency may from time to time specify, a sample of such quantity as the agency may consider adequate for any examination required to be made.

(X) If the certification agency so directs, the holder of the certificate shall not sell or offer for sale any lot in respect of which a sample is furnished under the preceding clause until the agency authorizes the sale of such lot.

(XI) The holder of the certificate shall on being directed by the certification agency that any part of a lot has been found by the said agency not to conform to prescribed standards of quality or purity specified by or under the Act, withdraw the remainder of that lot from sale and so far as may, in the particular circumstances of the case, be practicable, recall all issues already made from that lot.

(XII) The holder of the certificate shall comply with the provisions of the Act and these rules and with the directions given after nor less than one month's notice by the certification agency to such holder.

PART VIII- Appeals

18. The form and manner in which and the fee on payment of which the appeal may be referred :

1. Every memorandum of appeal under sub section (1) of section 11 shall be in writing and shall be accompanied by a copy of the decision of the certification agency against which it has been preferred and shall set fourth concisely and under distinct heads the grounds of objection to such decision without any argument, or narrative.
2. Every such memorandum of appeal shall be accompanied by a treasury receipt for a sum of Rs. 100/-.
3. Every such memorandum of appeal may be presented either in person or through an agent duly authorized in writing in this behalf by the appellant or may be sent by the registered post.

19. Procedure to be followed by the Appellate Authority : In deciding appeals under the Act the appellate authority shall exercise all the powers which a Court has and shall follow the same procedure which a Court follows in deciding appeals from the decree or order of an original Court under the Code of Civil procedure, 1908 (5 of 1908).

PART IX- Seed Analysis and Seed Inspectors

20. Qualifications of Seed Analysis : A person shall not be qualified for appointment as Seed Analyst unless he:-

(i) Possesses a Master's or equivalent degree in Agriculture or Agronomy or Botany or Horticulture of a University recognized for this purpose by the Government and has had not less than one year's experience in seed technology, or

(ii) Possesses a Bachelor's degree I Agriculture or Botany of a University recognized for this purpose by the Government and has had not less than three years experience in seed technology.

21. Duties of a Seed Analyst:

1. On receipt of a sample for analysis the Seed Analyst shall first ascertain that the mark and the seal or fastening as providing in clause (b) of sub section (1) of section 15 are intact and shall note the condition of the seals thereon.
2. The Seed Analyst analyzes the samples according to the provisions of the Act and these rules.
3. The Seed Analyst shall deliver the copy of the report of the result of the analysis to the persons specified in sub-section (1) of section 16.
4. The Seed Analyst shall from time to time forward to the State Government the reports giving the result of analytical work done by him.

22. Qualifications of Seed Inspectors: A person shall not be qualified for appointment as Seed Inspector unless he is a graduate in Agriculture of a University recognized for the purpose by the Government and has had not less than one years experience in seed production, or seed development or seed analysis or testing in a seed testing laboratory.

23. Duties of a Seed Inspector: In addition to the duties specified by the Act the Seed Inspector shall-

(a) Inspect as frequently as may be required by certification agency all places used for growing, storage or sale of any seed of any notified kind or variety.

(b) Satisfy himself that the conditions of the certificates are being observed;

(c) Procedure and send for analysis, if necessary, sample of any seeds, which he has reason to suspect are being produced stocked or sold or exhibited for sale in contravention of the provisions of the Act or these rules;

(d) Investigate any complaint, which may be made to him in writing in respect of any contravention of the provisions of the Act or these rules;

(e) Maintain a record of all inspections made and action taken by him in the performance of his duties including the taking of samples and the seizure of stocks and submit copies of such record to the Director of Agriculture or the certification agency as may be directed in this behalf;

(f) When so authorized by the State Government detain imported containers which he has reason to suspect contain seeds, import of which is prohibited except and in accordance with the provisions of the Act and these rules;

(g) Institute prosecutions in respect of breaches of the Act and these rules; and

(h) Perform such other duties as may be entrusted to him by the competent authority.

PART X- Sealing, Fastening, Dispatch and Analysis of Samples

24. Manner of taking Samples: Samples of any seed of any notified kind or variety for the purpose of analysis shall be taken in a clean dry container which shall be closed sufficiently tight to prevent leakage and entrance of moisture and shall be carefully sealed.

25. Containers to be labeled and addressed: All containers samples for analysis shall be properly labeled and the parcels shall be properly addressed. The label on any sample of seed sent for analysis shall bear:-

(a) Serial Number.

(b) Name of the sender with official designation, if any;

(c) Name of the person from whom the sample has been taken;

(d) Date and place of taking the sample;

(e) Kind or variety of the seed for analysis;

(f) Nature and quantity of preservative, if any, added to the sample;

26. Manner of Packing, Fastening and Sealing the Samples: All samples of seeds sent for analysis shall be packed, fastened and sealed in the following manner;

(a) The stopper shall first be securely fastened so as to prevent leakage of the containers in transit.

(b) The container shall than be completely wrapped in fairly strong thick paper. The ends of the paper shall be neatly folder in and affixed by means of gum or other adhesive.

(c) The paper cover shall be further secured by means of strong twine or thread both above and across the container and the twine or thread shall then be fastened on the paper cover by means of selling wax on which there shall be at least four distinct and clear impressions of the seal of the sender, of which one shall be at top of the clear impressions of the seal of sender, of which one shall be at the top of the packet, one at the bottom and the other two on the body of the packet. The knots of the twine or thread shall be covered by means of sealing wax bearing the impression of the seal sender.

27. From of Order: The order to be given in writing by the Seed Inspector under clause (C) of sub section (1) of section 14 shall be in Form III.

28. Form of Receipt for Records: When a seed inspector seizes any record, register, document or any other material object under clause (D) of sub section (1) of section 14, he shall issue a receipt in Form IV to the person concerned.

29. Samples how to be sent to the Seed Analyst: The container of sample for analysis shall be sent to the Seed Analyst by registered post or by hand in a sealed packet enclosed together with a memorandum in Form V in an outer cover addressed to the Seed Analyst.

30. Memorandum and Impression of seal to be sent separately: A copy of the memorandum and a specimen impression of the seal used to seal the packet shall be sent to the seed Analyst separately by registered post or delivered to him or to any person authorized by him.

31. Addition of preservatives to Samples: Any person taking a sample of seed for the purpose of analysis under the Act may add a preservative as may be specified from time to time to the sample for the purpose of maintaining it in a condition suitable for analysis.

32. Nature and Quantity of the Preservative to be noted on the Label: Whenever any preservative is added to a sample, the nature and quantity of the preservative added shall be clearly noted on the label to be affixed to the container.

33. Analysis of the Sample: On receipt of the packet, it shall be opened either by the Seed Analyst or by an officer authorized in writing in that behalf by the Seed Analyst, who shall record the condition of the seal on the packet. Analysis of the sample shall be carried out at the State Seed Laboratory in accordance with the procedure laid down by the central Government.

34. Form of Notice: The Notice to be given under clause (A) of sub section (1) of section 15 to the person from whom the seed inspector intends to take sample shall be in Form VI.

35. Form of Report: The report of the result of the analysis under sub section (1) or sub section (2) of section 16 shall be delivered or sent in Form VIII.

36. Fees: The fees payable in respect of the report from the Central Seed Laboratory under sub section (2) of section 16 shall be Rs. 10/- per sample of the seed analyzed.

37. Retaining of the Sample: The sample of any seed shall, under clause (C) of sub section (2) of section 15, be retained under a cool, dry environment to eliminate the loss of viability and insect proof or rat proof container. The Containers shall be dusted with suitable insecticides and the storage room fumigated to avoid infestation of samples by insects. The sample shall be packed in good quality containers of uniform shape and size before storage.

PART XI- Miscellaneous

38. Records : A person carrying on the business refereed to in section 7 shall maintain the following records, namely:-

(a) Stock record of seed;

(b) Record of the sale of seeds;

39. Form of Memorandum : The memorandum to be prepared under sub section (4) of section 14 shall be in Form VIII.

Seeds (control) order, 1983

GSR No. 932 (E), dated 30.12.1983: in exercise of the powers conferred by section 3 of the essential commodities Act, 1955 (10 of 1955), the central Government hereby makes the following order, namely:

Preliminary

1. **Short title and extent:** (1) this order may be called the seeds(control) order, 1983,
   (i) It extends to whole of India.
   (ii) It shall come into force on the 30th December, 1983.

This order is constitutionally valid, vide schedule 6, list 3, entry 37 and also within the powers of the central Government under section 3 of Essential commodities Act, 1955.

2. **Definition:** In this order, unless the context otherwise requires:
   (a) "Act" means the essential commodities Act, 1955 (10 of 1955);
   (b) "Controller" means a person appointed as controller of seeds by the central Government and includes any Person empowered by the controller under this order;
   (c) "Dealer' means a person carrying on the business of selling, exporting or importing seeds and includes an agent of a dealer;
   (d) "Export" means to taken or cause to be taken out from any place in India to a place outside India.
   (e) From" means a from appended to this order;

(f) "Import" means to bring or cause to be brought to any place in India from outside India;

(g) "Inspector" means an inspector of seeds appointed under clause 12;

(h) "Registering authority' means a licensing authority appointed under clause 11;

(i) "Seeds" means the seeds and defined in the seeds Act, 1966 (54 of 1966);

(j) "State Government' in relation to a Union Territory means the administrator thereof whatever designation known.

Dealer in seeds to be licensed

3. **Dealer to obtain licence**: (1) No person shall carry on the business of selling, exporting or importing seeds at any place except under and in accordance with the terms and conditions of licence granted to him under this order.

    (I) Notwithstanding anything contained in sub-clause (1), the state Government may, by notification in the official Gazette, exempt from the provisions of that sub-clause such class of dealers is such areas and subject to such conditions as may be specified in the notification.

4. **Application for licence**: Every person desiring to obtain a licence for selling, exporting or importing seeds shall make an application in duplicate in from 'A' together with a fee of rupees fifty for licence to licensing authority.

5. **Grant and refusal of licence**: (1) the licensing authority may, after making such enquiry as it thinks fit, grant a licence in form 'B' to any person who applies for it under clause 4:

Provided that a licence shall not be issued to a person,--

(a) Whose earlier licence granted under this order is made suspension, during the period of such suspension:

(b) Whose earlier licence granted under this order has been cancelled, within a period of one year from the date of such cancellation:

(c) Who has been convicted under the essential commodities Act, 1955 (10 of 1955) or any order issued there under within three years preceding the date of application.

2. When the licensing Authority Refuse to grant licence to a person who applies for it under clause 4, he shall recorded his reason for doing so.

6. **Period of validity of licence:** Every licence under this order, shall, unless previously or cancelled, remain valid for three years from the date of its issue.

7. **Renewal of licence:** (1) Every holder of licence desiring to renew the licence, shall, before the date of expiry of the licence, make an application for renewal in duplicate, to the licensing authority in form 'C' together with such fee, the licensing authority may renew the licence.

   (a) If any application for renewal is Nt made before the expiry of licence, but is made within one month from the date of additional fee of rupees twenty five, in addition to the fee for renewal of licence.

8. **Dealers to display stock and price list:** Every dealer of seeds shall display in this place of business,---

   (a) The opening and closing stocks, on daily basis, of different seeds held by him;

   (b) A list indicating prices or rates of different seeds.

9. **Dealers to give memorandum to purchaser:** Every dealer shall give a cash of credit memorandum to a purchaser of seeds.

10. **Power to distribute seeds:** where it is considered necessary to be so in public interest, the controller may, by an order in writing direct any producer or dealer to sell or distribute any seed in such manner as may be specified therein.

Enforcement Authority

11. **Appointment of licensing authority:** the state Government may by notification in the official Gazette appoint such number of persons as it thinks necessary the area within which each such license authority shall exercise his jurisdiction.

12. **Appointment of Inspectors:** the state Government may by notification in the official Gazette appoint such number of person as it thinks necessary to be Inspector and may in such notification define the local area within which each such Inspector shall exercise his jurisdiction.

13. **Inspector and Punishment:** (1) Any Inspector may with a view to securing compliance with this order;--

   (a) Require any dealer to give any information in his possession with respect to purchase storage and sale of seeds by him;

(b) Enter upon and search any permises where any seed is stored or provision of this order:

(c) Draw samples of seeds meant for sale. Export and seeds imported, send the same in accordance with the procedure laid down in schedule, to a laboratory notified under the conforms to standard of quality claimed:

(d) Seize or detain any seed in respect of which he has reason to believe that a contravention of this order has been committed or is being committed:

(e) Seize any books of accounts or document relating to any seed in respect of which he has reason to believe that a contravention of the order has been committed or is being committed.

Provided that the Inspector shall give a receipt, in respect of the books of accounts or documents seize, to the person from whom they have been seized:

Provided further that the seized books of accounts or documents shall be returned to the person from whom the same had been seized after copies thereof or extracts therefore, as certified by such person have been taken.

i. Subject to the provision of paragraph (d) of sub-clause (1) the provision, of section 100 of the code of criminal procedure, 1973 (2 of 1974) relating to search and seizure shall, so for as may be, apply to searches and seizures under this clause.

ii. Where any seed is seized by an Inspector under this clause, he shall forthwith report the fact of such seizure to a magistrate where upon the provision, 1973 (2 of 1974) shall, so far as may be, apply to the custody and disposal of such seed.

iii. Every person, if so required by an inspector, shall be bound to offer all necessary facilities to him for the purpose of enabling him to exercise his power under this clause.

14. **Time limit for analysis:** the laboratory to which a sample has been sent by an inspector for analysis under this order shall analyse the said samples and send the analysis report to the concerned inspector within 60 days from the date of receipt of the sample in the laboratory.

15. **Suspension/cancellation of licence:** the licensing authority may, after giving the holder of the licence an opportunity of being heard, suspend or cancel the licence on the following frounds, namely:

(a) That the licence had been obtained by misrepresentation as to a material particular: or

(b) That any of the provision of this order or any condition of licence has been contravened.

16. **Appeal**: Any person aggrieved by an order:--

(a) Refusing to grant, amend or renew the licence for sale export or import of seeds:

(b) Suspending or cancelling any licence.

May within sixty days from the date of the order, appeal to such authority as the state Government may specify in this behalf, and the decision of such authority shall be final:

Provided that on application for appeal shall accompany an appeal fee of rupees fifty.

Miscellaneous

17. **Amendment of licence:** the licence authority may, on receipt of a request in writing together with a fee of rupees ten from a dealer, amend the licence of such dealer.

18. **Maintenance of records and submission of returns etc.**: (1) Every dealer shall maintain such books, accounts and records relating to his business as directed by the state Government.

(I) Every dealer shall submit nonthly return relating to his business for the preceding month in form 'C' to the licensing authority by the day of every month.

EMELEM FORM-A
(See clause 4)

From of Application to obtain Dealer's License

To

The Registering Authority,
............(Place)
State of / U.T. of

1. Full name and address of the application.

(a) Name and postal address:

(b) Place of business (place give exact address):

(a) for sale:

(b) for storage:

2. Is it a proprietary/limited company/hindu undivided family concern? Give the Name(s) and address (as) of proprietor / partner/ manager / karta.
3. In what capacity this application is made;
   (i) proprietor :
   (ii) Partner:
   (iii) Manager:
   (iv) karta.
4. was the applicant ever convicted under the Essential Commodities Act, 1955 (10 of 1955), or any order issued there under during the three years preceding the date of application ? If so give details.
5. Give the details of seeds to be handled.
   S.No. ....Name of seed....................
6. I/ we have deposited the license fee of rupees fifty vide Challan No. .... dated ....in
7. Declaration;
   (a) I / we declare that the information given above is true to the best of my / our knowledge and belief and no part thereof is false.
   (b) I / we have carefully read the terms and conditions of the licence given in form 'B' appended to the seeds (control) order, 1983, and agree to abide by them.

Dated: Signature of Applicant

Place:

Note: Where the business of selling/exporting/importing seeds is intended to be carried on at more than one place, a separate license should be obtained for each such place.

For use in the office of licensing Authority

Date of receipt	Name and designation or officer receiving the application

license to carry on the business of a seeds

license No .... Date ....

subject to the provision of the seeds (control) order, 1983 and to the terms and conditions of this licence Shri/M/s. .... is hereby granted licence to sell, export / import and store for the said purposes of seeds.

(2) The licence shall carry on the aforesaid business at .... (place for storage and place for sale) .... (Tehsil or District).

Date: Licensing Authority

Seal: state of ....

T state of ....

TERMS AND CONDITIONS OF LICENCE

1. The licence shall be displayed at a prominent and conspicuous place in part of the business premises open to the public.
2. The holder of he licence shall comply with the provisions of the seeds (control) order, 1983 and the notifications issued there under and for the time being in force.
3. This licence comes into force with immediate effect and shall be valid upto unless previously cancelled or suspended.
4. The holder of the licence shall, from time to time, report to the licensing authority any change in the premises where he carries on his business of sale, export, import or storage for his said purposes of seeds.
5. The licence shall give every facility to the licensing authority or any purpose of inspecting his stock in any Authority for the purpose of inspecting his stock in any shop, depot or godown or other place / place used by him for the purpose of storage, sale or export of seeds.

EMBLEM

FORM-C

(See clause 7)

Application for Renewal of Licence to carry on the Business of a Deal in seeds.

To

The licence Authority,

............(place)

State of / U.T. of

I/ we hereby apply for renewal of the licence to carry on the business of dealer in seeds under the name and style of Shri/M/s .... The licence, desired to be renewed, was and allotted licence No. .... on the ......day of ....19 ....

Signature of Application(s)

Full name and address of the Application(s) ....

Date and place ....

Certified that the licence bearing No .... Granted on .... to carry on the business of a dealer in seeds at the premises situated .... Is hereby renewed upto the provisions of the seeds (control) order, 1983.

Licencing Authority

State of ....

Date:
Renewal No.
Seal:

FROM-D

(See clause 18)

1. Month and year

(Quantity in quintals)								
Crop variety	Class of seed	Opening stock on the first day of the month	Quantity purchased during the month	Total quantity imported	Total	Total quantity sold	Total quantity exported	Closing stock on the last day of the month
(1)	(2)	(3)	(4)	(5)	(6)	(7)	(7)	(9)
					(3+4+5)			6-(7+8)

2. Name and designation of licensee:

Signature of Dealer

Name:

Address:

SCHEDULE

(See clause)

Sealing, fastening Dispatch and Analysis of samples

(I) Manner of taking samples: samples of any seeds for the purposes of analysis shall be taken in a clean dry container which shall be closed sufficiently tight to prevent leakage and entrance of moisture and shall be carefully sealed.

(II) Sampling intensity: bulk sampling: when sampling seed lost are stored in bulk (heaps, bins, wagons, etc.) the following sampling intensity should be regarded as a minimum requirement for obtaining the "bulk sample"].

(a) up to 500 kilograms: At least 5 individual samples except in case of smaller number of samples is sufficient, not less than, 3 samples need be taken.

(b) 501 to 3000 kilograms: one individual sample for each 300 kilograms, but not less than 5 individual samples.

(c) 3001 to 20000 kilograms: one individual sample for each 500 kilograms but not less than to individual samples.

For seed in bulk the individual samples shall be distributed at random all over the bulk and the samples drawn from varying depths, bag samples—for seed lote in bage or other containers the following requirement:

(d) upto 5 containers: samples each container but always take at least 5 individual samples.

(e) Form 6 to 30 containers: samples at least one to every three containers but never less than 5.

(f) 31 containers or more: samples at least one in every five containers but never less than 10.

Unless doubts exits about the homogeneity of a lot, all such primary samples should be combined to make a composite sample of the lot for submitting to the seed testing laboratory for a heterogeneity test.

(III) Containers to be labeled and addressed: all containers containing samples for analysis shall be property labeled and the parcels shall be properly addressed. The label on any samples of seed sent for analysis shall bear,

(a) serial number:

(b) name of the sender with official designation, if any;

(c) name of the person from whom the samples has been taken;

(d) date and place of taking the sample;

(e) kind or variety of the seeds for analysis;

(f) nature and quantity of preservative, if any added to the sample.

(IV) Manner of packing, fastening and sealing the samples:-

All samples of seed sent for analysis shall be packed, fastened and sealed in the following manner:

(a) the stopper shall first on securely fastened so as to prevent leakage of the containers in transit;

(b) the container shall than be completely wrapped in fairy strong thick paper. The ends of the paper shall be meatly folder in and affixed by means of gums or other adhesive:

(c) the paper cover shall be further secured by means of strong twing or thread both above and across the container, and the twine or thread shall than be fastened on the paper cover by means of sealing was on which there shall be at least four distiner and clear impression of seal of the sender, of which, one shall be at the top of the packet, one at the bottom and the other two on the body of the packet. The knots of the twine or thread shall be covered by means of sealing was bearing the impressions of the seal of the sender.

Chapter - 14

Contract Farming

14.1 DEFINITION

Contract farming is generally defined as an agreement between farmers and processing and/or marketing firms for the production and supply of agricultural products under forward agreements, frequently at predetermined prices. Within this broad framework, there are different variants of contracts depending on the intensity of contractual arrangements.

Variations of contracts Market provision

- **Terms and conditions of future sale:** price, quality, quantity and timing etc.
- **Resource provision**: selected inputs, extension or credit including on occasions land preparation and technical advice
- **Management specifications**: The grower agrees to follow recommended production methods, inputs regime, and cultivation and harvesting specifications.

Variations of contracts

- **Market provision**: Terms and conditions of future sale: price, quality, quantity and timing etc.
- **Resource provision**: Selected inputs, extension or credit including on occasions land preparation and technical advice
- **Management specifications**: The grower agrees to follow recommended production methods, inputs regime, and cultivation and harvesting specifications.

Crops suitable for contract farming

- **Perishable**: cannot be stored for long periods and needs to find market immediately **bulky:** and therefore costlier to transport

- **Plantation crops:** growers cannot abandon the plantations or the estates and are locked into relationship with processor
- **Possible:** need for processing created inter-dependence between growers and processors, vulnerable to exploitation.
- **Variations in quality:** where crops vary in quality and quantity is important for processing
- **Unfamiliar:** medicinal plants like safed musli, ashwagandha etc. and new products for new markets like gherkins etc.

Models of contract farming

- Centralized model: Sponsor buys from a large number of small farmers, quota allocation and tight quality control. Suitable for tree crops. Annual crops, poultry, dairy etc. and products often requiring high degree of processing, such as tea or vegetables for canning or freezing.
- Nucleus estate model: Sponsor also manages a central estate or plantation. The central estate is usually used to guarantee throughout for the processing plant but is sometimes used only for research or breeding purposes. The sponsor provides significant amount of material and management inputs under the model.

Models of contract farming

- **The multipartite model** may involve a variety of organizations, frequently including statutory bodies, financial institutions

Individual entrepreneurs or small companies characterize the informal model

- **The intermediary model** involves sponsor in subcontracting linkages of farmers with intermediaries. There is risk of the sponsor losing control of production and quality as well as prices received by farmers.

Merits of C.F

- Access to modern technologies and support for small and marginal farmers (80%)
- Capital starved farmers cannot make major investment in land improvement and modern inputs
- Inputs, technical guidance, management skills, credit as well as knowledge of new technology.

- Pricing arrangement can significantly reduce the risk and uncertainty of market place
- Raise the yields of both contract and non-contract crops by improving the.
- Contributes to value addition by facilitating the emergence of agro-processing industry
- Better quality products, export market for small farmers

Risk factors in contract farming : Risks for farmers

- Particularly when growing new and unfamiliar crops, farmers face the risks of both market failure and production problems
- Contracted farmers may face land constraints due to a lack of security of tenure, thus jeopardizing sustainable long-term operations
- Social and cultural constraints may affect farmers' ability to produce to managers' specifications
- Poor management and lack of consultation with farmers may lead to farmer discontent
- Farmers may break the contract and sell the produce in alternative markets, sometimes encouraged by rival sponsors or ruling higher prices in open market.
- Farmers may divert inputs supplied on credit to other purposes, thereby reducing yields.

Risks for sponsors

- Contracted farmers may face land constraints due to a lack of security of tenure, thus jeopardizing sustainable long-term operations
- Social and cultural constraints may affect farmers' ability to produce to managers' specifications
- Poor management and lack of consultation with farmers may lead to farmer discontent
- Farmers may break the contract and sell the produce in alternative markets, sometimes encouraged by rival sponsors or ruling higher prices in open market.
- Farmers may divert inputs supplied on credit to other purposes, thereby reducing yields

Provisions of Model Act

- Model agreement format
- Registration of contracts
- Dispute redressal mechanism
- Indemnity of farmer's land
- Presently by Contract Act, not adequate.

Role of extension agencies

- Rights and liabilities of farmers
- Clauses of agreement
- Establishing linkage between farmers and sponsoring companies
- Crop diversification as per the agro-climatic conditions-extension
- Dispute redressal mechanism-awareness etc.
- Inefficient management or marketing problems can mean that quotas are manipulated so that all contracted production is not purchased
- Inefficient management can lead to overproduction, and in some cases sponsors may be tempted to manipulate quality standards in order to reduce purchases
- Sponsoring companies may be unreliable or exploit a monopoly position
- Indebtedness of farmers for production problems.

Contract Farming Status in different States

State	Crop	Company/ Corporate	Area (ha)
Karnataka	Ashwagandha Dhavana; Marigold & Caprica chili; Coleus; Gherkins Medicinal Plants	Himalaya Health Care, Mysore S.N.C. Oil Company; AVT Natural Products; Natural Remedies Private Ltd.; 20 Pvt. Companies*; Sami Labs Limited, Bangalore	8,350
Maharashtra	Soybean; Several fruits, vegetables, cereals, spices and pulses; Potato; Sugarcane, Orange	Tinna Oils and Chemicals; Ion Exchange Enviro Farms Ltd. (IEEFL); M/s Mahindra Sulabh; Cooperative Societies	134,800
Madhya Pradesh	Whet, Maize, Several fruits, vegetables, cereals, spices, pulses, Soyabean, Garlic and White onion	Cargil India; Ion Exchange; Enviro Farms; ITC-IBD; Mahindra Sulabh, Garlico Industries Limited.	1,200

Punjab	Tomato and Chilly; Barley, Basmati, Maize; Basmati, Groundnut, Potato and Tomato; Green vegetables and exotic vegetables	Nijjer Agro; United Breweries, Satnam overseas; Sukhjit Starch, Mahindra Shubhlabh Satnam Overseas, DD Intl. Amira Foods Pepsi Co, Punjab Agro Foods Park Limited.	100000
Tamil Nadu	Cotton, Maize, Paddy, Cotton, Marundu Koorkan (Medicinal Plant) (Coleus Forsk-holii), Maize, Gherkins	Super Spinning Mills; Bhuvi Care Pvt. Ltd.; Bhuvi Care Pvt. Ltd; Appachi Cotton Company; M/s Mahindra Sulabh	1,830
Chhatt-isgarh	Safed Musli; Tomato	Larson & Turbo; BEC Co.	
Uttaran-chal	Guar Gum	M/s Mahindra Sulabh	-
Haryana	Turmeric, Mentha, Su-nflower, White Musli	HAFED	-
Andhra Pradesh	White Viagra, Fruits, Vegetables and flowers, Gherkins, Cocoa, Oil palm	Nandan Farms, Aduri Natural products, ACE Agrotech, Mahendra, Cadbury, Godrej, Palm Tech, Simapuri Industry	23,000
Gujarat	Processing of Medici-nal Plants and Alovera	Reliance Group	
Orissa	Seeds (paddy, Ragi, Green gram, Arhar, ground nuts etc), Sugar, Eucalyptus	Orissa seeds Production Corporation, Shakti Sugar, J.K.Paper	7,200
Rajasthan	Exotic Vegetables	M/S Raj Tech Agro	8
West Bengal	Chip Quality Potato	Fritt-lay India	20

Contract Farming (Failure Cases)

State	Crop	Company/ Corporate	Districts	Reasons for failure
Rajasthan	Tomato (In the year 1997-98)	Hindustan Lever Limited	Jaipur (Bassi Area)	Market price being higher than the contract price.
Chhattisgarh	Kharif Maize (In the year 2002) in 568 hectare	Godrej Agrovet Co.	Mahasamund	The sponsoring company did not take interest in buying, so farmers sold the green cobs in the market instead of grain as per the agreement. Green cobs of maize have good market value in this area.

Chapter 15

Challenges Ahead

When it is the time of globalization, privatization and liberalization the challenges to seed industry will also increase. The challenges are such that the seed Industry will have to think over the seed production system and pattern.

The seed industry has to have light on cleaning, grading quality certification, storage, transportation and new issues. After globalization, export import of the commodity has to be produced of the international standards. Seed Industry has to think to produce the quality seed in such pattern that the final produce may be of International standards. They have to leave the traditional patterns. The Import and Export will have free trade within the countries, it may also be possible that there have guts in the market and the commodity produced may compete with the local producers or the local producer may not be benefited in the competition. Seed Industry has take those production, which will be produced by the farmer's must at least near International standards. The scientists have also to introduce varieties with the genetic characters coinciding to the international standards. Same way Ministry of Agriculture will also have to think of the challenges of these type ahead.

After WTO agreement the free trade within the member countries will come into force. The WTO refers the several specification

1. Free Movement: Export and Import within the countries.
2. Restricting movement.
3. Price stability.
4. Traceblity.
5. Standardization and homonisation of the product.

Seed Industry has to maintain the above standard to be in the competition and stability.

Chapter - 16

Directorates of State Seed Certification Agencies

S.N.	Addresses	Fax No.	Telephone No.
1.	Director Gujarat State Seed Certification Agency Beej Pramanan Bhawan. Near Shyamal Raw House, Opp. Gurukul Raw House, AHEMDABAD – 380015	079-26734116	079-26734110 26734116 26779063
2.	Director Maharashtra State Seed Certification Agency Shastri Nagar, Akola-444004	0724-2258712	0724-2258713
3.	Director Karnataka State Seed Cert. Agency. KAIC Premises Opp. Baptist Hospital Hebbel, Bangalore-560024	080-23415506	080-23415505 23415652
4.	Managing Director MP State Seeds Cert. Agency office Complex, Gautam Nagar, Bhopal.	0755-2583903	0755-2583629
5.	Director Orissa State Seed Cert. Agency Govt. of Orissa Bhubaneswar – 751003	0674-2562078	0674-2561783 2392935
6.	Director Punjab State Seed Cert. Agency SCO No.-38, Sector –22-A CHANDIGARH	0172-2706494	0172-2706494
7.	Director Uttranchal State Seed Cert. Agency. DEHRADUN	0135-2760734	0135-2760861
8.	Director Assam State Seed Certification Agency, Ram Kirshana Mission Rd., Ulubar, GUWAHATI-781006		0361-2478805
9.	Director AP State Seed Cert. Agency, HACA Bhawan opp. Public Garden HYDERABAD-500004	040-23231021	040-23237016
10.	Director Rajasthan State Seed & Organic Production Cert. Agency, G.B. pant Krishi Bhawan Bhagwan Das Road JAIPUR	0141-2227456	0141-2227104
11.	Divisional Seed Cert. Officer, Talab Tillo JAMMU		091-2505201

12.	Director W.B. State Seeds Cert. Agency, Govt. of West Bengal Writer's Building KOLKATA-700001	033-2451636	033-22145856
13.	Director UP State Seed Cert. Agency Horticulture Complex Kariyappa marg, Alam Bagh, LUCKNOW	0522-2451636	0522-2451639 2841013
14.	Director Hariyana State Seed Cert. Agency Bay No.-11-12, Sector-14, PANCHLULA (HARYANA)	0172-2567642	0172-2570662
15.	Director Bihar State Seed Cert. Agency Mithapur Farm, Mithapur PATNA		0612-2354358
16.	Managing Director Chhattisgarh State Seed Cert. Agency Labhandi Farm Krishak Nagar RAIPUR – 1	0771-2442319	0771-2442319
17.	Director H.P. State Seeds Cert. Agency, Nalagarh House SHIMLA – 171005	0177-2830643	0177-2830643
18	Chief Seed Certification Officer, Seed Certification Agency Govt.of NCT Khaiberpan, Civil Lines New Delhi-54		011-23819707
19	Director Deptt of Seed Certification PO Box No 5403 Kavundapalayam Post TTC Office Building Complex Mettupalayam Post Coimbtoor	0422-2457554	0422-2432984

Chapter - 17

List of State Seed Corporations Limited

S.N.	Addresses	Fax No.	Telephone No.
1.	Managing Director Maharashtra State Seeds Corp. Ltd. Mahabeej Bhawan Amravati Road, P.B. No. 119 AKOLA-444004	0724-2558655	0724-2258402 2259116 2258161
2.	Managing Director Karnataka State Seeds corpn. Ltd. Bellary Road, Hebbal BANGALORE-560024	080-23415895	080-23415652 23415656
3.	Managing Director MP Rajya Beej Evam Farm Vikas Nigam Mother Teresa Marg, Arera Hills BHOPAL - 462016	0755-2775048	0755-2775043
4.	Managing Director Orissa State Seeds corpn. Ltd. Santrapur BHUBANESWAR-751002	0764-2340096	0764-2340573
5.	Managing Director Punjab State Seeds Corpn. Ltd. SCO No. 835-36, Sector 22-A GANDHINAGAR-160022		0172-2701582
6.	Managing Director Gujarat State Seeds Corpn. Ltd."Beej Bhawan" Sector-10A, GANDHINAGAR-382043	079-23248094	079-23221384
7.	Managing Director Assam Seeds Corporation Ltd. Khanapura GUWAHATI - 781022	0361-2330529	0361-2338472
8.	Managing Director A.P. State Seeds Development Corporation Ltd. 5-10-193 (2nd floor) HACA Bhawan Opp. Public Gardens HYDERABAD-500004	040-23243401	040-23240099 23236088
9.	Managing Director Rajasthan State Seeds Corporation Ltd. G.B. Pant Krishi Bhawan JAIPUR	0141-5116231	0141-5116231
10.	Managing Director W.B. State Seeds Corporation Ltd. 4 Gangadhar Babu Lane KOLKATA-700012	033-22375591	033-22367859 22259311

11.	Managing Director U.P. Seeds Development Corporation Ltd. C-973-974 B. Faizabad Road Mahanagar LUCKNOW	0522-2384845	0522-2384802 2385356 2385832
12.	Managing Director US & TDC Ltd. Pant Nagar, P.O. Haldi Distt. NAINITAL (UTTRANCHAL)	05944-230351 233830	05944-230263 260222
13.	Managing Director Haryana Seeds Development Corporation Ltd. Bay No. 2-3, Sector-2 PANCHKULA(HR)	0172-2577583	0172-2577582
14.	Managing Director Bihar Rajya Beej Nigam Ltd. Mithapur Farm Campus, Mithapur PATNA		0612-2344396
15.	Managing Director Chhattisgarha Rajya Beej Nigam Ltd. New Krishi Upaj mandi Complex Pandritrai RAIPUR-492001	0771-4046039	0771-4046039

Chapter - 18

Codes of States and their registered Seed Processing Plants Bihar State Seed Certification Agency

क्र. सं.	संयंत्र का नाम व पता
1	बिहार सीड्स कॉर्पोरेशन, बेतिया, प.च.
2	बिहार राज्य बीज निगम लिमिटेड, कुदरा, भभुआ
3	प्रभारी पदाधिकारी, तिरहुत कृषि महाविधालय, ढोली, मुजफ्फूरपुर
4	मसीना बीज प्राईवेट लिमिटेड, ताजपुर रोड, समस्तीपुर
5	नायक सीड्स, भोरेजयराम, समस्तीपुर
6	मंगलगढ़ बीज निगम, मंगलगढ़, समस्तीपुर
7	कृषि बीज विकास निगम, मसीना, ताजपुर रोड, समस्तीपुर
8	विश्वनाथ चौधरी, हरिहरपुर खेडी, मसीना, समस्तीपुर
9	चक्र मसीना बीज निगम, चक्का खानपुर, समस्तीपुर
10	त्रिवेणी सीड्स, शाहपुर, भोरेजयराम, समस्तीपुर
11	मसीना बीज फार्म, हरिहरपुर खेडी, मसीना, समस्तीपुर
12	प्रमाणित मसीना बीज प्राइवेट लिमिटेड, मिल्की, मसीना, समस्तीपुर
13	मसीना बीज प्राईवेट लिमिटेड, मुक्तापुर, समस्तीपुर
14	ए. आर. एम. लेबोरेट्री, ताजपुर रोड, समस्तीपुर
15	सदानन्दपुर बीज फार्म, सदानन्दपुर, बेगूसराय
16	गजेन्द्र झा, इलमासनगर रोड, मथुरापुर, समस्तीपुर
17	प्राचार्य, बिहार कृषि महाविद्यालय, सबौर, भागलपुर
18	उत्पादन प्रमुख, सोन कमाण्ड बाजार समिति, बिहटा, पटना
21	प्रशिक्षण संगठन, कृषि विज्ञान केन्द्र, जापानी फार्म, आरा, भोजपुर
22	सौरभ सीड्स कम्पनी, भ.आई. पी. कालनी, भभुआ
23	पैग्रोइन्डट्रीज, बोडीहा कला, वजीरगंज, गया
24	सत्या सीड्स, ग्राम पो. हरदिया, भोजपुर
25	प्रभात कुमार, पुरानी बाजार, लखिसराय
26	क्षेत्रीय प्रबन्धक, राष्ट्रीय बीज निगम लिमिटेड, शेखपुरा, पटना
27	खुबीलाल एग्रो केमिकल प्रा.लि., डाक बंगला रोड, खाजी मोहल्ला, बिहारशरीफ, नालन्दा

क्र. सं.	संयंत्र का नाग व पता
28	अनमोल हाइब्रीड सीड्स, चैनपुर, भभुआ
29	आर.एन.सीड्स फार्म, ग्राम हिरामन बिगहा, जहानाबाद
30	प्रभारी पदाधिकारी, सिंचाई अनुसंधान केन्द्र, ब्रिगगंज, रोहतास
31	भारत सीड्स फार्म, नूरसराय, ग्राम पो. कौडी, मेयार, नालन्दा
32	सोन गंगा सीड्स, इन्डस्ट्रियल एरिया, बिक्रमगंज, रोहतास
33	सोना सीड्स, लालपुर, कुदरा, कोमुर
34	कृष्णा सीड्स, बंका, सिकन्दरपुर, मुजफ्फर
35	क्षेत्रीय प्रबन्धक, राष्ट्रीय बीज निगम लिमिटेड, शेखपुरा, पटना

Maharashtra State Seed Certification Agency, Akola

1	Seed Processing Plant MSSC, Shioni, Akola
2	SPP & Ginning Centre MSSC Shioni, Akola
3	Seed Processing Plant Vegetable, Shioni, Akola
4	HOD Dr. PDKV Agriculture, Eng. SPP, Akola
5	Chief Seed Production Officer, Dr. PDKV (SPP) Wani Rambhapur, Akola
6	Chief Seed Production Officer, Dr. PDKV (Ginning Center) Wani Rambhapur, Akola
7	Ginning Center, MSSC, Patur, Akola
8	The New Akot Cotton Ginning And Pressing Co. Ltd., Akot, Akola
9	SPP, MSSC, Washim
10	SPP Mahalaxmi Seed Co. M.A.I.D.C., Akola
11	SPP Anup Seed Co. Kurum, Akola
12	Basant Seed Kanheri, Akola
13	Mahesh Hy. Seed Co.
14	Nuzividu Seeds Co. Akola
15	Tirupati Seeds Co. Malegaon, Washim
16	Anupam Agro Industries M.A.I.DC., Akola
17	Poonam Seeds Co., Washim
18	Shrikant Cotton Industries G.I.P. Unit Shioda, Akola
19	Balaji Co-Op Sugar Ind. Masalapen, Washim
20	Gauri Seeds & Bio Tech Corporation Chikhalgaon, Akola
21	Janki Seeds & Research Pvt. Limited, Akola
22	Nishant Sahakari Gramin Patpurawartha Sanstha, Chohotta Bajar, Akola
23	SPP, MSSC, Tapowan, Amaraoti
24	National Seeds Corporation, NSC, Amaraoti
25	Gramin Krishi SPP Asegaon Purna, Amaraoti
26	Amol Seeds Co. Chandikapur, Amaraoti
27	Shiv Agro, Waknath, Tq, Dhamangaon, Amaraoti
28	Shree Seeds Co. Amaraoti
29	Maharashtra Agro Seeds Bhatkuli Amaraoti
30	Mahalaxmi Seeds Co. Masod, Amaraoti
31	Hindustan Agro Servises SPP Badnera, Amaraoti
32	Ravi Seeds Co., Amaraoti
33	Daryapur Sahakari Gining and Pressing Factory (MSSC) Amaraoti
34	Vidharbh Agro Seeds Co. Walgaon, Amaraoti
35	The Vidharbha Co.-op Soc. Ltd., Maphed MSSC, Amaraoti
36	JDS Ginning Sategaon Tq Anjangaon, MSSC, Amaraoti

37 Malani Ginning & SPP, Yeotmal
38 Libarti Seeds Co. Yeotmal
39 SPP Butibori, Nagpur
40 Ankur Seeds Co., Nagpur
41 SPP MSSC Shelu, Wardha
42 Yeshoda Hy. Seeds Co., Kumbhi, Wardha
43 SPP MSSC Mul, Chandrapur
44 SPP, MSSC Gadeaon, Bhandara
45 Daftari Agro Pvt. LTd., Shelu, Wardha
46 Dwarka Seeds, Bajargaon, Nagpur
47 Nuzevedu, Seeds Pvt. Ltd. Gadchiroli
48 Purti Agro Tech SPP Dhapewada, Nagpur
49 Arun Hy. Seeds Pvt. LTd. Rotha, Wardha
50 SPP, MSSC, Chikhali, Buldhana
51 SPP, MSSC, Malkapur, Buldhana
51 SPP, MSSC, Khamgaon, Buldhana
52 Ankur Seeds Pvt. Ltd., Malkapur, Buldhana
53 Mahyco Seeds Ltd., Dhanora, Buldhana
54 Golden Seeds Co. Chikhali, Buldhana
55 Akshya Seeds SPP Malgani Tq. Chikhali Buldhana
56 Balaji SPP, Shindi Harali, Tq. Chikhali, Buldhana
57 Harit Kranti Seeds Processor, Deolgaon Raja, Buldhana
58 Sheetal Seeds, SPP, Deolgaon Raja, Buldhana
59 Birla Agro Pvt. LTd., Khamgaon, MSSC, Buldhana
60 Buldhana Urban Seed, Chandaknagar, Dongarkhandal, Buldhana
61 SPP MSSC, Parbhani
62 The Sahayog Shakari Kharadi Vikri Sangh Ltd., Parbhani
63 Mararthwada Krishi Vidhyapeeth SPP, Parbhani
64 Balaji Seeds Co. Parbhani
65 Balaji Hy. Seeds Corporation, Nanded
66 Ganesh Seeds Company, Nanded
67 SPP MSSC, Cotton, Dhanegaon, Nanded
68 Maruti Seeds Company, Nanded
69 Parsewar Seeds Co. Nanded
70 Samarth Seeds Pvt. Limited, Nanded
71 Laxmi Hybrid Seeds Co. Ltd., Nanded
72 Santosh Seeds Company, Nanded
73 SPP MSSC Dhoki, Osmanabad
74 SPP DMO Godwon, Parbhani
75 Kharedi Vikri Sangh Godwon, Hingoli
76 SPP Texcom MIDC, Nanded
77 Anil Agro Agencies, Jawalabajar, Hingoli
78 SPP MSSC, Erandol, Jalgaon
79 Ginning Center, Pachora, Jalgaon
80 Arihant Ginning And Pressing Factory Dongar Kathora, Jalgaon
81 National Seeds Corporation, Jalgaon
82 Bhagashri Agro Inputs, Jalgaon
83 Kohinoor eeds Pvt. Ltd., Pachora, Jalgaon
84 SPP Agriculture College, Dhule
85 SPP MSSC, Satana, Nashik
86 National Horticulture Research & Development, Nashik
87 Loknayak Jayprakash Narayan Shakari Soot Girni, Untawad, Nandurbar
88 Shri Salasar Cotton Industries, Pachora, Jalgaon
89 Sahayog SPP, Tandulwadi Tq, Rawer, Jalgaon
90 SPP Seed Multiplication Farm, Musalgaon, Nashik
91 National Seed Corporation, Aurangabad
92 Green Gold Seeds Ltd, Aurangabad

93 Ajit Seeds Pvt. LTd., Chitegaon, Aurangabad

94 Krishidhan Seeds Jalna-1

95 Krishidhan Seeds Pvt. Ltd., Jalna-2

96 SPP MSSC, Jalna

97 Anand Ginning & Preesing Ind. Pathrod, Beed

98 Mulchand Fulchand Krishi Uyog Ltd., Jalna

99 Jaybhavani Seeds & Research Ltd Partur, Jalna

100 Jindal Traders, SPP, Jalna

101 SPP Khandala , Shrirampur, A. Nagar

102 SPP Market Yard Shrirampur, A. Nagar

103 Mahatma Fule Krishi Vidhyapeeth, Rahuri, A. Nagar

104 Mauli Agro Servises, Kopergaon, A. Nagar

105 Superintendent, Agriculture College, Manjri, Pune

106 Solapur District Milk Federation, Tembhurni, Solapur

107 Shivamrut Milk Federation, Akluj, Solapur

108 National Seeds Corporation, Sangli

109 SPP MSSC, Ashta, Sangli

110 Sheti Padwidhar Sahakari Sangh Ltd., Kolhapur

111 Dr. Babasheb Sawat Kokan Krishi Vidhyapeeth, SPP Wakawali, Dapoli

112 Deepak Agro Solutions Ltd. Wagholi, Pune

113 Hy. Seeds Co. Dudhediya, Shrirampur, A. Nagar

114 Shiva Seeds, Rahuri, A. Nagar

Karnataka State Seed Certification Agency

1 National Seeds Corporation Ltd., Hebbal, Bangalore 24

2 National Seeds Corporation Ltd, Maghacvi Building, Near Bus Stand, Haveri

3 Karnataka State Seeds Corporation Ltd., P.B. Road, Dharwad 580 005

4 Karnataka State Seeds Corporation Ltd., P.B. Road, Bellary 583 101

5 Karnataka State Seeds Corporation Ltd., PWD Camp, Shindhanur, Raichur

6 Varada Seeds Corporation Near bus Stand Haveri 581116

7 Karnataka State Seeds Corporation LTd., APMC Yard, Haveri, 581110

8 Sri Ramakrishna Seeds Processing Unit, No. 21, Godown, APMC Yard, Siraguppa, Bellary District

9 Karnataka State Seeds Corporation Ltd., Near Jadalethimmanahalli Cross, Agaiagurki (P), Chikkabaliapur-562101

10 Karnataka State Seeds Corporation Ltd., Tekal Road, Kolar-563101

11 Mahanandi Seeds Karryapally, Muragamalla Road, Chitamani, Kolar Dist.

12 University of Agriculture Science

13 Agriculture College Campus, Dharwad-580005 Phone 52068

14 Karnataka State Seeds Corporation Ltd., Kotnur, Gulbarga Dist

15 Sri Raghavendra Seed Processing Unit , Near Market Yard Rani Ganj, Raichur

16 Vinayaka Seeds & Agriculture Development Corporation, 88-A, R.M.C. Road, Devangore-577003

17 Veerabhadreswara Agro Traders, P.b. Road, Renebenuur-581115 Haveri Dist

18 Karnataka State Seeds Corporation Ltd., Hinakal, Mysore

19 Karnataka State Seeds Corporation Ltd., Vidyagiri, Bagalkot-587102

20 Karnataka Janatha Hybrid Seed Near bus Stand, Siddlaghatta, Kolar Dist

21 Regional Oil Seeds, Crowers Co-operative Societies Union Ltd., P.B. No. 21, Hospet, Bellary Dist.

22 Krishi Kuteera Seeds Processing Unit, Ankalikars Kiran Collage Sindhagi Road, Bijapur

23 Karnataka State Seeds Corporation Ltd., APMC Yard Channarayapatna

24 Agriculture Centre & Service Association U.B. Hills Dharwad

25 Sagar Seed Processing Unit, C/o Fertilizer & Agriculture Traders, Bagalkote-587101

26 Ravikiran Enterprises, D013/14, Banumatah Circle, Mysore-570024

27 Varada Seeds Farm City Press, Godown, Near Sangam Thearte, 16th Ward, Bellary

28 Somanath Seed Processing Unit, KSSDC Industrial Estate, Special Plot No. 10, Hatalageri Naka Masori Gadag

29 National Seed Corporation Ltd., Doddamandiganahalli, Kandli Post Hassan, 573201

30 The Central State Farm, SFCI Ltd., Jawalagere Sindhanur, Tq., Raichur Dist.

31 Karnataka State Seeds Corporation Ltd., K.R. Nagar, Mysore dist

32 Karnataka State Seeds Corporation Ltd., RAPCMS, Rice Mill Complex, Mandya

33 Karnataka State Seeds Corporation Ltd., Holehonnur Road, Gurupara, Shimoga

34 Bhadra Hyvrid Seed APMC Yard, Davanagere-577002

35 University of Agriculture Science RMS, Kathalagere, Channagiri Taluk, Shimoga

36 University of Agriculture Science NSP Block, J Block, GKVK, Bangalore

37 Karnataka Agro Seeds H. Nehru Road, Behind Golgembaz Road,

38 Karnataka State Seeds corporation Ltd., Laxmeswara Road., Gadag-582101

39 Karnataka State Seeds Corporation Ltd. Industrial Estate, Bhalki, Bidai Dist

40 Karnataka State Seeds corporation Ltd. Girinagar, Kyatha Sandra, tumkur

41 Karnataka State Seeds Corporation Ltd., UAS, Campus Raichur

42 National Seeds Co. Ltd., 732/A, guggarabhatti Bangalore Road, Bellary

43 Ganga Seed Processing Unit Shed No/ 17, 18 & 20 Industrial Estate Raichur

44 Sri Marula Siddeswara Seed Processing Unit Hulihalli, Ranebennur, Tq. Haveri Dist.

45 Karnataka State Seeds Corporation Ltd., UAS, Complex, Bheemarayangudi, UKP Cada, Shahapur, Gulbarga Dist.

46 Anta Seeds Processing Unit, Sanganakal Bellary Dist.

47 Karnataka State Seeds Corporation Limited, Koppal

48 Raju Seed Processing Unit, C/o Raj Trading Company Shed No. C-9-10, Industrial Estate Raichur, 584101

49 The Vegatable Seeds Processing Unit, KSDH, Lalbagh c/o DDH Office, Horticulture Department

50 Krishi Pragathi Seed Processing Unit, Deveragudda Road, Ranebennur Tq. Haveri Dist

51 Varada Seed Farm, No. 929-A, A3/ABCD Ward XVIII, Near Sales Tax Check Post Ananthpur Road, Bellary

52 Karnataka Hi Tech Agro Enterprises Gaddikere industries

53 Sri Ram Rajeshwari Seeds Processing Unit, Opp. Karwallal Ginnign fact. Vth Wards, Siraguppa, Bellary Dist

54 Maharastra Hybrid Seeds Co. Ltd., Kamdod NH-4, Ranebennur

55 Suraj Seed Processing Unit Godown No. 7, Cancle Market, Malapur Dharwad

56 Sri Banasharkari Seed Processing Unit D.No. 114, RMC Link Road, Davangore

57 Bangalore Milk Union Ltd., Seed Processing Unit Rajankunte Bangalore North

58 Mahesh Hybrid Seeds Processing Plant, Opp St. Joseph School, Chickkaballapur

59 Royal Seeds Processing Unit Moka Road, Sanganakallu Bellary

60 Sri Mrityunjaya Seeds Processing Unit, Kumhapur, Belgaum Road, Nerendra, Dharwad

61 Ganga Kaveri Seeds Processing Unit, Industrial Area, Mysore Road, Pandavapura, 571434

62 The Karnataka Hybrid Seed Corporation Seed Processing Unit, P.B. Road, Raneberpur, Haveri Dist.

63 Basavehwara Seeds Behind Tahasildar Office, Lingasugar-584122,

64 Patil Seed Processing Unit, R.S. No. 438, Deveragudda Road, Kunabev Cross

65 Surya Seed Processing Unit, C-14, Industrial Estate Bangalore Road, Bellary

66 A.G. sun Seeds Pvt. Ltd., Plot No. 32, 2nd Stage, Mundaragi Industrial Area, Bellary

67 Simhadri Agro Seeds Processing Unit, Devaragudi Pali, Mallasandra Post, Bagepalla, Tq. Kolar Dist.

68 Arunodaya Seeds Processing Unit, Chelor Road, Chitamani - 563125 Kolor dist

69 Indo American Hybrid Seeds (P.) Ltd., Seeds Processing Unit Medleri Road, Near Railway Gate, Ranebennur, Haven Dist

70 Sri Anjaneya Gro Tech Pvt. Ltd., Seed Unit, 376/7, P.B. Road, Davangore

71 Central Poddar Seed Production Farm, Government of India, Ministry of Agriculture, Development of Animal Husbandary and Dairying Hesaraghatta, Bangalore-88

72 Laxmi Seeds And Farms, Near Wind Power, Mulgund Road, Gadag-582 103

Divisional Seed Certification Officer, Department of Agriculture, Lal Mandi, Srinagar

1 Seed Processing Plant, R.O. Jawahar Nagar, Srinagar 190 008

2 Seed Processing Plant, Malangpura, Agri Farm, District Pulwama

Office of the Divisional Seed Certification Officer, Agriculture Department, Jammu

1 Major Seed Processing Plant No. 1, 1977-78

2 Major Seed Processing Plant No. 2 1994-95

3 Vegetable Seed Processing Plant, No. 3 2001-02

छत्तीसगढ़ राज्य बीज प्रमाणीकरण संस्था, इंदिरा गांधी कृषि विश्वविद्यालय परिसर, कृषक नगर, रायपुर

1 छत्तीसगढ़ राज्य बीज एवम कृषि विकास निगम लिमिटेड, कांकेर

2 इंदिरा गाधी कृषि विश्वविद्यालय, रायपुर

3 इंदिरा गाधी कृषि विश्वविद्यालय, रायपुर

4 फेयर फूड प्राइवेट लिमिटेड, सकरी, बिलासपुर

5 महामाया सीड्स कार्पोरेशन, अजिरमा, अम्बिकापुर

6 इंदिरा गांधी कृषि विश्वविद्यालय, अम्बिकापुर

7 एस.बी. एग्रो सीड्स, अम्बिकापुर

8 कृषि विभाग, रायपुर

9 कृषि विभाग, दुर्ग

10 कृषि विभाग, धमतरी

11 उद्यान विभाग, रायपुर

12 कृषि विभाग, रायगढ़

13 कृषि विभाग, जगदलपुर

14 मैसर्स बम्लेश्वरी सीड्स, सहदेव नगर, राजनांदगांव, राजनांदगांव

15 कृषि विभाग, कोरिया

16 राष्ट्रीय बीज परियोजना, इंन्दिरा गांधी कृषि विश्वविद्यालय, रायपुर

17 कृषि विभाग, अम्बिकापुर

18 छत्तीसगढ़ राज्य बीज एवम कृषि विकास निगम लिमिटेड, बिलासपुर

19 इंदिरा गांधी कृषि विश्वविद्यालय, रायपुर

20 छत्तीसगढ़ राज्य बीज एवम कृषि विकास निगम लिमिटेड, अम्बिकापुर

21 छत्तीसगढ़ राज्य बीज एवम कृषि विकास निगम लिमिटेड, रायपुर

22 छत्तीसगढ़ राज्य बीज एवम कृषि विकास निगम लिमिटेड, महासमुंद

23 छत्तीसगढ़ राज्य बीज एवम कृषि विकास निगम लिमिटेड, रायगढ़

24 छत्तीसगढ़ राज्य बीज एवम कृषि विकास निगम लिमिटेड, धमतरी

25 छत्तीसगढ़ राज्य बीज एवम कृषि विकास निगम लिमिटेड, राजनांदगांव

26 छत्तीसगढ़ राज्य बीज एवम कृषि विकास निगम लिमिटेड, जगदलपुर

27 छत्तीसगढ़ राज्य बीज एवम कृषि विकास निगम लिमिटेड, कांकेर

28 छत्तीसगढ़ राज्य बीज एवम कृषि विकास निगम लिमिटेड, जांजगीर

29 छत्तीसगढ़ राज्य बीज एवम कृषि विकास निगम लिमिटेड, दुर्ग

30 छत्तीसगढ़ राज्य बीज एवम कृषि विकास निगम लिमिटेड, कवर्धों

31 कृषि विभाग, जांजगीर

32 कृषि विभाग, राजनांदगावं

33 कृषि विभाग, रायगढ़

34 कृषि विभाग, रायपुर

35 कृषि विभाग, जगदलपुर

36 आकाश सीड्स एण्ड कम्पनी, अम्बिकापुर

List of Registered Seed Processing Plants, Grading, Sealing centres in Himachal Pradesh

1 Department of Agriculture, Bilaspur

2 Department of Agriculture, Chamba

3 Department of Agriculture, Hamirpur

4 Dept. of Agriculture, Jassur, Kangra

5 Dept. of Agriculture, Bhatton, Kangra

6 CSK, HPKVV, Palampur, Kangra

7 Private, Yangthang, Kinnaur

8 Department of Agriculture, Kullu

9 IARI, Katrain, Kullu

10 Private, Klath, Kullu

11 LPS, Chauribihal, Kullu

12 LPS, Bahang, Kullu

13 LPS, Manali, Kullu

14 LPS, Ani Chowai, Kullu

15 Department of Agriculture, Spiti Lahoul

16 Dept. of Agriculture, Bhangrotu, Mandi

17 Private, Bhuira, Mandi

18 Private, Dhura, Shimla

19 Private, Tikkar, Devrighat Khadrala, Shimla

20 Private, Theog, Fagu, Shilaroo, Shimla

21 Private, Kharapather, Shimla

22 Private, Rohru, Badiyara, Chirgaon

23 Dept. of Agriculture, Dhaulakuan, Sirmour

24 Dept. of Agriculture, Paonta Sahib Sirmour

25 CSK HPKVV, Dhaulakuan, Sirmour

26 NSC, Solan, Solan

27 Dept. of Agriculture, Manjholi, Solan

28 Dept. of Agriculture, Pekhubella, Una,

29 Department of Agriculture, Rajgarh, Una
30 Dept. of Agriculture, Thianbagh, Sirmour
31 Dept. of Agriculture, Rajgundha, Kangra

Haryana State Seed Certification Agency

Processing Plants

1 HSDC, Umri, Kurukshetra
2 HSDC, M-5, Indst. Area, Yamuna Nagar
3 HSDC, 4th K.M. Stone, Sirsa Road, Hisar
4 HSDC, Hisar Road, Hisar
5 NSC, 15, IDC Delhi Road, Hisar
6 NSC, Hansi Road, Karnal
7 HAU, Old Ludas Road, Hisar
8 CSF, Central Colony, at Farm
9 IFFOCO, Plot No. 61, Sector-28, Industrial Area, Hisar
10 IFFCO, Sunder Cold Store, G.T. Road, Karnal
11 Sehwag Agri Farm Chhudani, VPO Chhudani, Rohtak
12 Haryana Seed Farm, VPO Mangali, Hisar
13 Field Grading, Karnal Zone
14 Field Grading Hisar Zone
15 Field Grading Sirsa Zone
16 Super Seeds Pvt. Ltd., Talwandi Rana, Hisar
17 Shiv Ganga Hybrid Seeds, Ols Ludas Road, Hisar
18 Sharda Enterprises, Near Petrol Pump, Sirsa Road, Fatehabad
19 Surya Seeds and Chemicals, Ist K.M. Stone, Bhuna Road, Fatehabad
20 Sehrawat Agri Farm, V & PO Bhanguri, Palwal
21 Nirakari Seed Co., Chauhan Nagar, Dabwali Road, Sirsa
22 Kuber Seds, Sirsa Road, Fatehabad
23 Golden Seed Corpration & Chemical, Ahmedpur Road, Industrial Area, Sirsa
24 Durga Seed Co. Plot No 2-B, Iind Industrial Area Sirsa
25 Predial Seed Pvt. Ltd., 5th K.M. Stone, Sirsa Road, Fatehbad
26 Haryana Seed Regd. 27th, IIIrd Industrial Area, Sirsa
27 Parkash Seeds, Near Bharam Vidhalya, Hostel Road, Hisar
28 HSDC, Dangra Road, Tohana
29 Surya Seed & Chemical, Ist K.M. Stone, Bhuna Road, Fatehabad
30 C.S.F. Hisar Thaska Colony, C.S.F. Hisar
31 Hargopal Harkishan, 6th K.M. Stone, Barwala Road, Hisar
32 Unnat Beej Co., Opp Air Force Sattion, Dabwali Road, Mirpur Sirsa
33 Unnat Seed Corporation, Opp. Police Station, Samalkha, Dist Panipat
34 Sun Seed Corp. 29/5, Industrial Area, Fatehabad, Hisar
35 Shaktivardhak Hybrid Seed Ltd., 29/5, Industrial Area, Fatehabad, Hisar
36 Uttam Seeds, Sirsa Road, Daya Nand Rishi Vihar Colony, Hisar
37 Vardhman Seeds, Railway Road, Kurukshetra
38 Kisan Seed Farm, Plot No. 79, 2nd. Industrial Area, Sirsa
39 Shiv Ganga Hybrid Seeds, Old Ludas Road, Hisar
40 KRIBHCO, 7th K.M. Stone, Barwala Road, Hisar
41 Goodwill Hybrid Seed, Uklana, Distt. Hisar
42 Supreme Seeds, Hajrawa Road, Fatehabad
43 Bajaj Seed Pvt. Sirsa Road, Fatehabad
44 Punjan Seed Co., 54, 2nd Industrial Area, Sirsa
45 Dhan Luxmi Seeds,, Dabwali Road, Sirsa
46 Chetak Seeds, 11th K.M. Stone, barwala Road, Hisar
47 Parag Seeds Pvt. Ltd., 18, Indust. Area, Fatehabad

48 Kanav Seeds, Shamshabad, Aligarh Road, Palwal

49 Sahib Seed Pvt. Ltd., Vijay Nagar, Karnal

50 HSDC, Narhera Road, Pataudi

51 Parbhat Seed Traders, 49-50, Industrial Area, Sector-2, Pipli, Kurukshetra

52 SSPL, 8th K.M. Stone, Barwala Road, Talwandi Rana, Hisar

53 K.D. Seeds, VPO Mathana, Dist. Kurusksheta

54 United Rice Land, VPO Samanabahu, Karnal

55 Thaneshwar Co-Operative Society, Amin Road, Kurukshetra

56 Shri Ram Seeds, Main Road, Siwan, Dist Kaithal

57 Naman Seed, Near Village Dobh, Dist. Rohtak

58 Narindera Hybrid Seed Co., Near FCI Godown M. Dabwali Sirsa

59 Lakeshya Seed, Narwana Road, Palwan, Dist Jind

60 Modern Seed Corp. Village Rasulpur Indri, Dist Karnal

61 Volga Seed, Opp. Dera Sacha Sauda, Rania Dist Sirsa

62 Jagdish Hybrid Seed Co., Jind Road Hansi

63 Ish Seed Corporation, Talwara Road, Ellenabad, Dist Sirsa

64 Paras Seed Co., Sirsa

65 Kadiyan Seed Corp. Vill. Lalani Sultan Pur, Tarouri, Dist Karnal

66 Pankaj Seed Farm, Gangwa Road, Hisar

67 Grow Tech Seed P. Ltd., Ghora Farm Road, Hisar

68 Kurukshetra Seed P. Ltd., Subash Mandi, Kurukshetra

69 Shree Balaji Seed, Main Road, Siwan, Dist. Kaithal

70 Quality Hybrid Seed, Village Mangali, Dist. Hisar

71 Sanjeevni Seed Co. Satya Nagar, Hisar

72 SSPL, 8th K.M. Stone, Barwala Road, Talwandi Rana, Hisar

73 Sawstik Seed Co., Near Dhani Satnam Singh Rania

74 PICRIC Ltd., 56-57, G.T. Road, Larsauli, Dist Sonepat

75 PGF Ltd., VPO Pattikaikalyan, Samalakha, Dist Panipat

76 Sahankar Seeds, Vill. Anta, Safidon, Dist. Jind

77 Haryana Seed Co., Chirao, Dist. Karnal

78 A-One Agro Products VPO Beer, Near Chandan Nagar, Hisar

79 Shan Seed, Gudha

80 Haryana Beej Co., Dabawali Road, Sirsa

81 Mohinder Seed Corp. Hodel Faridabad

82 Shiva Seed Co., Jind Road Barawala

83 Jagjeet Seed Processing Unit, Shamshabad Patti Sirsa

84 Bhatti Seed Farm, Shergarh, Dabwali,

85 Naveen Seed, Indri Dist Karnal

86 Juneja Seed Processing Unit, Jatpura Road, Kachawa, Karnal

87 RDS Farm, HAU, Hisar

88 Metro Hybrid Seed, Rohtak

89 Kamboj Export, Indergarh, Dist. Karnal

90 Anand Hybrid Seeds Co. Mirzapur Road, Hisar

91 Shiv Shakti Seed, Dabwali Road, Sirsa

92 Arjun Seed Padhana

93 Hygine Kamboj Seed Farm, Indergarh, Indri, Dist Karnal

94 Deepak Seeds, Mandi Adampur

95 Dhaliwal Seed Pvt. Ltd, Pipli Road, Near Central Jail, Kurukshetra,

96 Ashok Hybrid Seed Hisar, D.N. College Road, Satya Nagar, Hisar

97 Parag Seed Ltd, 4th K.M.Stone, G.T. Road, Fatehabad

98 American Hybrid Seed Corpo. Ratia Chungi, Fatehabad

99 Rajan Seed Chamanpura

100 Hare Krishna Seed Jhansa Road, VPO Dhurala Dist. Kurukshetra

101 Saraswati Seed, Main Road Siwan Dist Kathail

102 Sunstar Overseas Ltd., 40th, K.M. Stone, G.T. Road, Bahalgarh, Sonipat

103 Hisar Seed Corporation, VPO Gorota, Tehsil Hodel, Dist Faridabad

104 Punjab Agro Seed, Industrial Area, Sirsa

105 Shaktiman Seed, Dabwali Road, Sirsa

106 Aman Seed Co. Kharar Alipur, Hisar

107 Shiv Ganga Hybrid Seed Pvt. Ltd. Old Ludas Road, Hisar

108 Neel Kanth Seed Co., Market Committee, Gaushala Road, Karnal

109 Rajpat Agriculture Store, Village Sahapur, Hisar

110 Skyfin Green Land Ltd., Village Khora Kheri, Ghurunda, Dist Karnal

111 Semen's Seed Co. Dhramgarh Old Road, Hansi Road, Dhramgarh Dist. Panipat

112 Krishi Pooja Seed VPO Gidha, Indri Dist Karnal

113 Ankit Hybrid Seed Co., Dabwali Road, Industrial Area, Sirsa

114 NHRDF, VPO Salarum Dist Karnal

115 Dhaliwal Agro, Jhansa Road, Village Chhiba

116 Panchmukhi Seeds, 7th K.M. Rewari Road, Silana, Dist Jajjhar

117 Om Agri Seeds, VPO Narata Indri.

118 Mukta Seed Co. Kelnia Road, S.B. Patti, Dist. Sirsa

119 Jai Bharat Seed Co. 6th K.M. Stone Mirzapur Road, Hisar

120 Krishan Hybrid Seed Chandan Nagar

121 Harshana Beej Co., B village Hidayatpur, Palwal, Dist Faridabad

122 Biochem Organics Pvt. Ltd, 10th K.M. Stone Barwala Road, Jugalan, Hisar

123 Diamond Seed, 1-K.M. Stone, Bhiwani Road, Hansi, Dist Hisar

124 Hari Bhumi Seed, Chulan More, Ismaila, Dist. Rohtak

125 Jagrati bio Tech Hybrid Seeds, Village Uklana

126 Shaktivaradhak Hybrid Seeds P. Ltd. Ludas Road, Arya Nagar, Hisar

127 Durga Dass Hybrid Seed Pvt. Ltd., Village Ganuali, Dist. Panchkula

128 Agricare Seed, Village Samargopalpur Rohtak

129 K.P. Agro, Bigher Road, Fatehabad

130 Evergreen Seed Co, 10th, K.M Rajgarh Road, Hisar

131 Bhole Ram Kailash Kumar, P.No. 90 Sector 27-28, Hisar

132 New Hisar Beej Co., Mandi Adampur, Hisar

133 Bhumi Seed Company, VPO Kuttabad, Sirsa

134 V.K. Seeds Village Nahera P.O. Khera Indri Karnal

135 Krishak Dhan Seed, Railway Road, Charkhi Dadri Bhiwani

136 Sadbhawna Seeds, VPO Kussar, Rania Distric Sirsa

Seed Producers whose Seed Processing Plants are Registered with Punjab State Seed Certification Authority

1 Punjab State Sees Corporation Ltd., Kishangarh Road, Kartarpur, Dist Jalandhar

2 Punjab State Sees Corporation Ltd., A-12 Industrial Area, Focal Point Ludhiana

3 Aujla Seed, Shop No. 54, New Grain Market, Machhiwara, Dist Ludhiana

4 Punjab State Sees Corporation Ltd., Industrial Focal Point Kotkapura, Dist Faridkot

5 Punjab State Sees Corporation Ltd., Dayal Nagar, Abohar, Dist Ferozpur

6 National Seed Corporation, C-25, Focal Point, Jalandhar

7 Director, University Seed Farm, Ladhowal, Ludhiana

8 PAU Regional Research Station, Gurdaspur

9 PAU regional Rice Research Station, Kapurthala

10 PAU Seed Farm, Vill Nariangarh, PO Amloh Dist. Fatehgarh Sahib

11 PAU Regional Research Station, Kheri, Dist Sangrur

12 PAU Raja Harinder Singh Research Station, Vill. Bir Sikhanwala, Dist Faridkot

13 Randhawa Seed Farm, VPO Talwandi Lal Singh Teh Batla, Dist Gurdaspur

14 Assiciate Director, Training, PAU Krishi Vigyan, Kender Kapurthala

15 Sahu Agro Centre, Seed & Pesticidies, Shop No. 2, Opp. PAU Gate No.1, Ferozpur Road, Ludhiana

16 Garh Seed Corporation 7-G, Sarabha Nagar, Opp PAU Gate No. 2, Ferozpur Road, Ludhiana

17 Punjab Agri Seed Farm, Gaushala Road, Mansa

18 Raghbir Agro Enterprises Pvt. Ltd, 42 Col, Raghvir Marg, Bhupinder Nagar, Patiala

19 Akal Seed Farm, Gurudwala Mastuana Sahib Mastuana, Dist Sangrur

20 Surinder Seed Farm 23, New Grain Market Muktsar

21 Punjab Beej Company, Opp Grain Market Samrala Road, Machhiwara Dist Ludhiana

22 Potato Chandigarh Jurisdiction

23 Potato Govt. Seed Farm, Jalandhar Jurisdiction

24 Potato Private Producers, Jalandhar Jurisdication

25 Potato Govt Farms, Ludhiana Jurisdiction

26 Potato Private Producers, Ludhiana Jurisdiction

27 PAU Cotton Research Station, Abohar Dist. Ferozepur

28 Milkfed Seed Processing Plant Post Box 2, Bassi Pathana, Dist. Fatehgarh Sahib

29 Punjab Agriculture Seed Farm, Opp. PAU Gate No. 2, Ferozpur Road, Ludhiana

30 Punjab Beej Bhandar, Gidderbaha, Dist Mukatsar

31 S. Amrinder Singh, Punjab Agriculture Farm. V-Mahalan, PO Mahlan Jhallian, Dist Ropar

32 Jain Agricultural Farm, Old Bazar Opp. Jain Sabha, Sardulgarh, Dist. Mansa

33 Partap Agricultural Farm, vill. Jaito Sarja, The Batala Dist Gurdaspur

34 Plantsman Seed, C/o National Nursary, Rajbaha Road, Patiala

35 Anmol Seed Corporation, H.No. 38, Model Town Phsae-I, Jandiala Guru, Dist. Amritsar

36 Goodwill Seeds, 42 Raghuvir Marg, Bhupindera Nagar, Patiala

37 Dhola Seed Corporation, Grain Market, Tappa Dist Sangrur

38 Hindustan Seed Co., Sanghara Road, Barnala, Dist Sangrur

39 Matta Seed Farm, D-43, Focal Point Kotkapura

40 Proline Seed Co. India Pvt. Ltd Jalalabad East Dist Ferozpur

41 National Seed Corporation Ltd., A-2, Focal Point Dabwali Road, Bhatinda

42 Technocare Seed Agencies SCF-43-A, New Grain Market Mukatsar

43 Sardar Hybrid Seed Co., Raman Mandi, Dist Sangrur

44 Punjab Hybrid Seed Co. Patiala Road, Sunam Dist Sangrur

45 Krishi Vigyan Kendra, Rauni, Patiala

46 Hara Seed And Nursary Near Rauru Sahib Gurdwara GT Road, Sahnewal Dist. Ludhiana

47 Surinder Pal Singh & Co. 5-G, Sarabha Nagar, Ludhiana

48 Gold Field Seed Pvt. Ltd, B-XX-3073, Gurdev Nagar, Malhar Palace, Ludhiana

49 D.S. Agro Seeds, Old Anj Mandi Nabha Gate, Patiala

50 Bright Seed Farm, V-Khawaspur, P.O. Malakpur, Dist. Ropar
51 Capital Seed & Food Products, Shop No. 31 Grain Market, Ropar
52 Grow Rice Seed And Nursaries, VPO Pannam, Teh. Garhshankar, Dist Hoshiarpur
53 Pawan Kumar Naresh Kumar Talwandi Sabo Road, Mour Mandi. Dist Bhatinda
54 Pacca Seeds, Near Mahabir cotton Fatory, Raman Mandi Dist. Bhatinda
55 MYCO Seed Farm, Behind Thakur Mohalla Gidderbaha, Dist Mukastsar
56 Ladhowal Seeds, Ladhowal, Dist Ludhiana
57 Laxmi Seed Farm, C-66, Focal Point Kotkapura, Dist Faridkot
58 Hara Seeds, Opp. PAU Gate No. 1, Ludhaiana
59 Seed Farm Development Company, Fazilka Road, Jalalabad West Dist. Ferozpur
60 Sarv Seed Company Dharma Singhwala Moga
61 Shri Durga Seed Store, Gidderbaha, Dist Muktsar
62 Krishak Bharti Co.-Op. Ltd., KRIBHCO Seed Processing Unit VPO Killianwali, Abohar, Dist Ferozpur
63 Sartaj Agriculture Seed, Vill Nangal Sohal PO Ramdass, Dist. Amritsar
64 Kissan Seed Corporation Satta Bazar, Gidderbaha, Dist Muktsar
65 Kohinoor Seed Field Indioa Pvt. Ltd Grain Market, Mour Mandi Distt. Bathinda
66 Shri Ganesh Seed Farm, Main Bazar, Mour Mandi, Distt. Bhatinda
67 Gulzar Seed VPO Bazipur Moga Road, Ferozpur
68 GEE Agri Seed Kharyal Road, Sunam, Distt. Sangrur
69 Thind Seed Farm, Vill Jodha Nagri, PO Dehriwal, The Baba Bakala Dist. Amritsar

Seed Processing Plant Registration, West Bangal

1 Pallishri Ltd., Arambag, Hooghly
2 Pallishri Ltd., Arambag, Hooghly
3 Pallishri Ltd., Arambag, Hooghly
5 Pallishri Ltd., Arambag, Hooghly
6 Pallishri Ltd., Arambag, Hooghly
7 Pallishri Ltd., Arambag, Hooghly
8 Pallishri Ltd., Arambag, Hooghly
9 Pallishri Ltd., Arambag, Hooghly
10 Scientist In Charge Central Seed Research Station For Jute & Allied Fibres, PO Bud, Burdwan
11 Pallishri Ltd., Arambag, Hooghly
13 Pallishri Ltd., Arambag, Hooghly
14 Birbhum Chemicals & Fertiligers Ichhapur, Illambazar, Birbhum
15 Birbhum Chemicals & Fertiligers Ichhapur, Illambazar, Birbhum
16 Birbhum Chemicals & Fertiligers Ichhapur, Illambazar, Birbhum
17 Birbhum Chemicals & Fertiligers Ichhapur, Illambazar, Birbhum
18 Birbhum Chemicals & Fertiligers Ichhapur, Illambazar, Birbhum
19 Birbhum Chemicals & Fertiligers Ichhapur, Illambazar, Birbhum
20 Birbhum Chemicals & Fertiligers Ichhapur, Illambazar, Birbhum
21 Annapurna Seed, Fatehpur
22 Pallishri Ltd., Arambag, Hooghly
23 Birbhum Chemical & Fertiliger Ichhapur, Illumbazar, Birbhum
24 Birbhum Chemical & Fertiliger Ichhapur, Illumbazar, Birbhum
25 Birbhum Chemical & Fertiliger Ichhapur, Illumbazar, Birbhum
26 Pallishree Ltd., 2nd Unit, Ramgarh, Simlapal, Bankura
27 Pallishree Ltd., 2nd Unit, Ramgarh, Simlapal, Bankura

28 Pallishree Ltd., 2nd Unit, Ramgarh, Simlapal, Bankura

29 Pallishree Ltd., 2nd Unit, Ramgarh, Simlapal, Bankura

30 Pallishree Ltd., 2nd Unit, Ramgarh, Simlapal, Bankura

31 Pallishree Ltd., 2nd Unit, Ramgarh, Simlapal, Bankura

32 Pallishree Ltd., 2nd Unit, Ramgarh, Simlapal, Bankura

33 Pallishree Ltd., 2nd Unit, Ramgarh, Simlapal, Bankura

34 Pallishree Ltd., 2nd Unit, Ramgarh, Simlapal, Bankura

35 Pallishree Ltd., 2nd Unit, Ramgarh, Simlapal, Bankura

36 Tirupati Agro Seed Distributors

38 Annapurna Seed, Fatehpur

40 Hall Mark Industries, Shyam

41 Birbhum Chemicals & Firtiliger Illambazar, Birbhum

42 Pallishree Ltd., Arambag, Hooghly

43 Pallishree Ltd., Arambag, Hooghly

44 Pallishree Ltd., Arambag, Hooghly

45 Tirupati Agro Seed Dristributer

46 Arambag Seed Unit, Aarambag, Hoogh

47 Lakshi Sagar Seed Unit, Bankura

48 Lakshi Sagar Seed Unit, Bankura

49 Hall Mark Industries at Shyam Nagar Kantabagau Mouza Under P.S. Joypur Bankura

50 Mali Agro Tech Pvt. Ltd Vill Jymarsatp PO Ranaghat Dt. Nadia Jl No. 157 Plc No. 16217

51 West Bengal State Seed Farm Hargram, Bhatar, Burdwan

52 Ramnagar Seed Farm Hargram, Bhatar, Burdwan

53 Ramnagar Seed Farm Hargram, Bhatar, Burdwan

54 Gontra S.K.U.S. Ltd, Vill Gontra PO ghetugachhi, Dist Nadia

55 Anapurna Seed, Vill Daichanda PO Huria Mouza Kamalpur PS Khand Ghosh Dust Burdwan

56 Anapurna Seed, Vill Daichanda PO Huria Mouza Kamalpur PS Khand Ghosh Dust Burdwan

57 Anapurna Seed, Vill Daichanda PO Huria Mouza Kamalpur PS Khand Ghosh Dust Burdwan

58 Anapurna Seed, Vill Daichanda PO Huria Mouza Kamalpur, PS Khand Ghosh Dust Burdwan

59 Anapurna Seed, Vill Daichanda PO Huria Mouza Kamalpur, PS Khand Ghosh Dust Burdwan

60 Sudarshan Seed Vill Makurgrar PO Makurgram, P/s Bankura Mouza Bon

61 Arambad Seed Farm, Vill PO Uchalan, PS Uchalan, Plot No. 4200,

62 GMS Agro Enterprises Vill PO Gidhni, PS Jambani, JL No. 113, Khatian No 234/2, Mouza Kherejora

63 Bhumishree Seed Farm Gojdharpara, PS Berhampur Dist Murshidabad

64 Annapurna Seed Vill Daichanda, PO Huria, PS Khandi Ghosh, Dist Burdwan

65 Annapurna Seed Vill Daichanda, PO Huria, PS Khandi Ghosh, Dist Burdwan

66 Annapurna Seed Vill Daichanda, PO Huria, PS Khandi Ghosh, Dist Burdwan

67 Annapurna Seed Vill Daichanda, PO Huria, PS Khandi Ghosh, Dist Burdwan

68 Hallmark Industries Pvt. Ltd., Vill Shyam Nagar, PS Joypur Mouza Kantabon

69 West Bengal State Seed 21, KNC Road, Barasat, 24, Parganas

70 Regional Station For Forage Production & Demonstration, Kalyani PO Netaji Subhash Sanatorium Kalyani

71 GMS Agritech Pvt Ltd, 1st Floor, BH-93, Sec-II, Salt Lake City

72 Vishal Seed Corporatio, Vill & PS Barabazar, PO Barabhum

73 Krishi Sabayan & Company Shibaniwas More, PS Krishnagan 38, No. Songhata Mouza, Khatian

74 Director of Farm, BCKV Mohanpur Teching Farm, Mondouri, PS Haringhata PO Mohanpur

Madhya Pradesh State Seed Certification Agency, Bhopal

1 Samrat Seeds, Balodi, Betul

2 M.P. State Seed and Farm Development Corporation, Betul

3 Khandelwal Agro Agency, Betul

4 Govt. Agri Farm, Gudgaon, Betul

5 Bio Bliss Beej Utpadak Sah. Samiti, Betul

6 Surya Beej Utpadak Sah. Samiti, Hidli Betul

7 Maha Asha Agro Foods, Bhesdehi Betul

8 M.P. State Seed & Farm Development Corporation, Govindpura Bhopal

9 National Seed Corporation, Govindpura Bhopal

10 Abhinav Seed, Rasla Khedi Bhopal

11 Shanti Seeds Bhanpur, Bhopal

12 Vindhya Seeds, Mubarakpur, Bhopal

13 Haridhan Seeds, Charkheda Harda

14 M.P. State Seed and Farm Development Corporation, Harda

15 Uday Seeds, Sodalpur Harda

16 Shriram Beej Utpadak Sahkari Samiti, Bedi Harda

17 Surya Seeds and Agriculture, Harda

18 M.P. State Seeds & Farm Development Corporation, Seoni Malva,Hoshangabad

19 Agriculture Research Center, JNKVV, Power Kheda, Hoshangabad

20 M.P. State Seeds & Farm Development Corporation, Babai Hoshangabad

21 M.P. Uddhog Vikas Nigam, Babai, Hoshangabad

22 Narmada Beej Utpadak Swa. Sahkari Samiti, Babai, Hoshangabad

23 Horticulture & Farm Forestry, Silari Hoshangabad

24 Monika Seeds Private Limited, Pipariya Hoshangabad

25 Gaurav Seeds, Dhandiya Khaparkheda, Hoshangabad

26 M.P. Oil Fed, Banapura, Hoshangabad

27 Vikalp Agritech, Shobhapur, Hoshangabad

28 Sanwariya Seeds, Shivpur, Hoshangabad

29 Shrinath Seeds, Hoshangabad

30 Ravi Seeds, Khaparkheda, Hoshangabad

31 M.P. State Seeds & Farm Development Corporation, Begumganj, Raisen

32 M.P. State Seeds & Farm Development Corporation, Udaypura, Raisen

33 M.P. State Seeds & Farm Development Corporation, Bareli, Raisen

34 Osha Beej Utpadak Sah. Samiti, Kychwara, Raisen

35 M.P. State Seeds & Farm Development Corporation, Biora, Rajgarh

36 M.P. State Seeds & Farm Development Corporation, Jeerapur, Rajgarh

37 Shri Krishna Seeds, Jeerapur, Rajgarh

38 Pankah Seeds, Machalpur, Rajgarh

39 Sankalp Seeds, Machalpur, Rajgarh

40 Ravi Seeds & Research Biora, Rajgarh

41 Gopal Seed Agriculture, Kurawar Rajgarh

42 Govt. Agri Farm, Fanda Sehore

43 M.P. Oil Fed, Amlaha Sehore

44 M.P. Oil Fed, Pachama Sehore

45 Global Seeds, Jharkheda Sehore

46 Shri Ram Genetics Bhikhedi Sehore

47 JNKVV, Sehore

48 M.P. State Seeds and Farm Development Corporation, Vidisha
49 Premier India Seeds, Mandi Bamora, Vidisha
50 Betwanchal Seeds, Vidisha
51 Kashtkar Seeds, Lashkarpur, Vidisha
52 Mangal Saain Seeds, Vidisha
53 Kribhco Seeds, Balgarh, Dewas
54 Sundaram Seeds & Protien, Dewas
55 Indo Agrigenetics, Dewas
56 M.P. State Seeds & Farm Development Corporation, Khategaon, Dewas
57 M.P. State Seeds & Farm Development Corporation, Hat Pipaliya Dewas
58 Green Gold Agritech Durgapur, Dewas
59 Bhaskar Seeds, Sirolya Seeds, Dewas
60 Adhyaksha Harnawada Beej, Dewas
61 Amisha Beej Utpadak Sah. Samiti, Bhada Pipalya, Dewas
62 Govt. Agri. Farm, Chandra Sekhar, Dewas
63 M.P. State Seeds & Farm Development Corporation, Dhar
64 M.P. Oil Fed, Dhar
65 Amisha Seeds, Manawar, Dhar
66 Nutan Quality Seeds, Manawar, Dhar
67 Rama & Rama Seeds, Kapsi, Dhar
68 Rallis Hybrid Pipal Kheda, Dhar
69 Shankar Hybrid Seeds, Bhamnod, Dhar
70 M.P. State Seeds & Farm Development Corporation, Manawar, Dhar
71 Bhumika Seeds, Khalghat, Dhar
72 M.P. State Seed & Farm Development Corporation, Badnawar, Dhar
73 Govt. Agri Farm, Dhar
74 Indo Germ Seed Company, Khalghat, Dhar
75 National Seed Corporation, Indore
76 Sagar Agri Private Ltd., Indore
77 Soyabean Processor Association of India, Indore
78 Onila Seeds, Indore
79 Krishidhan Seed Private Ltd, Indore
80 Laxmi Seeds, Indore
81 Mahadhan Seeds Pvt. Ltd., Indore
82 Vigour Seeds, Indore
83 Pawan Seeds, Indore
84 Eagle Seeds, Kshipra, Indore
85 A.S.N. Agrigenetics, Rauu, Indore
86 Eagle Seeds And bio, Rakhodi, Indore
87 Mohara Seeds, Nraval, Indore
88 Mayur Seeds, Palda, Indore
89 M.P. Oil Fed, Indore
90 Natioan Bagwani Anushandhan Kendra, Indore
91 Mamta Beej Utpadak Sah. Samiti, Budhaniya Hatod, Indore
92 Hiz Seeds, Indore
93 Krutika Seeds, Indore
94 Manika Seeds, Indore
95 Mayur Seeds, Sanwer, Indore
96 Katya Beej Utppadak Sah. Samiti, Indore
97 Sardar Beej Samiti, Bachhoda, Indore
98 Varsha Seeds, Hatod, Indore
99 S.K. Seeds Corporation, Indore
100 Bisco Seeds Tech Private Ltd, Devas Naka, Indore
101 M.P. State Seeds & Farm Development corporation, Thandla, Jhabua
102 Govt. Agri. Farm, Jobat, Jhabua
103 Vijay Laxmi Hybrid Seeds, Bamaniya Jhabua
104 M.P. Progressive Seeds, Bamaniya, Jhabua
105 Satyam Hybrid Seeds, Ojhar, Badwani
106 Kundan Seeds, Anjad, Badwani
107 Heera Beej Company, Badwani

108 Sona Seeds, Badwani

109 M.P. State Seeds & Farm Development Corporation, Khandwa

110 Maharashtra Hybrid Seeds, Khandwa

111 Sethi Seeds, Khandwa

112 Khandwa Oils, Siltiya Pandhani, Khandwa

113 College of Agriculture, Khandwa

114 Amit Seeds, Khandwa

115 Shri Hari Seeds and Agro, Haidarpur

116 Siddharth Seeds, Sirpur, Khandwa

117 Swastik Beej Utpadak Samiti, Khargoan

118 M.P. State Seeds and Farm Development Corporation (for Groundnut), Khargoan

119 M.P. State Seeds & Farm Development Corporation, Khargoan

120 M.P. State Seeds & Farm Development Corporation, Badwah, Khargoan

121 Narmada Beej Utpadak Kray Awam Vikray Samiti, Khargoan

122 Om Seeds, Badwani, Khargoan

123 Navkar Agritech, Khargoan

124 Sanawad Hybrid Seeds, Sanawad, Khargoan

125 Maa Ambika Beej Utpadak Samiti, Khargoan

126 M.P. State Seeds & Farm Development Corporation, Khargoan

127 M.P. Agro Fertilizers, Mandsaur, Mandsaur

128 Sipani Anusandhan Farm, Changali, Mandsaur

129 Oswal Seeds and Chamical Ltd., Neemuch

130 Ratan Raj Seeds, Jaora, Ratlam

131 Sagar Agro Input, Alote, Ratlam

132 Ravi Seeds and Research, Alote, Ratlam

133 Shri Parshvanath Agro, Ratlam

134 Narmada Agro, Alote, Ratlam

135 M.P. State Seeds and Farm Development Corporation, Shajapur, Shjajapur

136 Bhoomiputra Breej, Kiloda, Shajapur

137 Monika Seeds, Mohan Barodiya Shajapur

138 Krishi Dev Seeds, Aagar, Malwa, Shajapur

139 Monisha Seeds, Akodiya Mandi, Shajapur

140 Dev Shri Seeds, Sangimana, Shajapur

141 Kamdenu Biotech Seeds Khokra Kala, Shajapur

142 Dev Narayan Seeds, Agar Malva, Shajapur

143 M.P. State Seeds and Farm Development Corporation, Ujjain

144 M.P. Oil Fed, Nagjhiri, Ujjain

145 Pragya Seeds, Mahidpur, Ujjain

146 Sagar Agri Input, Pandya Khedi, Ujjain

147 Arihant Agro Seeds, Chak, Kamed, Ujjain

148 Navlakha Seeds, Mahidpur, Ujjain

149 Kisan Ware Housing Sah. Samiti, Ujjain

150 Vikram Beej Utpadak Sah. Samiti Ltd.,

151 Vasundhara Bio Organization, Ujjain

152 Mahalaxmi Beej Utpadak Sah. Samiti Ltd, Baldi Badnagar, Ujjain

153 Deepak Beej Utpadak Awam Vipnan Sah. Samiti

154 Vikram Seeds, Narvar, Ujjain

155 Krishi Dev Seeds, Narvar, Ujjain

156 M.P. State Seeds and Farm Development Corporation, Chhatarpur

157 Beej Utpadak Sah. Samiti, Lugasi, Chhatarpur

158 Unnat Krishak Beej Utpadak Sah. Samiti, Chhatarpur

159 M.P. State Seeds & Farm Development Corporation, Damoh

160 Jai Kisham Beej Utpadak Samiti, Kuwarpur, Damoh

161 Jwala Seeds, Pipariya Damoh

162 M.P. State Seeds & Farm Development Corporation, Janakpur Panna, Panna

163 Kisan Beej Utpadak Sahkari Samiti, Sundara, Panna

164 Deputy Director Agriculture, Panna,

165 M.P. State Seeds & Farm Development Corporation, Sagar, Sagar

166 M.P. State Seeds & Farm Development Corporation, Deori, Sagar

167 Bundelkhand Seeds, Chandpur Rehli, Sagar

168 J.N.K.V.V., Rajaua Sagar, Sagar

169 Beej Utpadak Sah. Samiti, Besa Kala Sagar, Sagar

170 Annapurna Beej Utpadak Sah. Samiti, Amarmau Sagar, Sagar

171 Vishal Beej Prakriya Kendra, Sindgua, Sagar

172 Sanmati Beej Utpadak Sah. Samiti, Gaurjhamar, Sagar

173 Krashak Beej Uptadak Sah. Samiti, Nivodiya, Sagar,

174 Uttam Beej Utpadak Sah. Samiti, Luharra, Sagar

175 Bhaskar Beej Utpadak Sahkari Samiti Narwa, Shahgarh, Sagar

176 M.P. State Seeds & Farm Development Corporation, Kundeshwar, Tikamgarh

177 J.N.K.V.V., Tikamgarh

178 National Seeds Corporation, Niwari, Tikamgarh

179 Raj Seeds, Khargapur, Tikamgarh

180 Bundelkhand Sah. Samiti, Alampur, Tikamgarh

181 Shriram Raja Beej Utpadak Sah. Samiti, Niwari, Tikamgarh

182 Krishana Beej Utpadak Sah. Samiti, Niwari, Tikamgarh

183 College of Agriculture, Tikamgarh

184 M.P. State Seeds & Farm Development Corporation, Garra, Balaghat

185 Krishi vistar Awam Prashikshan Kendra, Baraseoni, Balaghat

186 M.P. State Seeds & Farm Development Corporation, Chhindwara

187 J.N.K.V.V., Chandangaon

188 M.P.Oil Fed., Chhindwara

189 National Seeds Corporation, Chhindwara

190 Govt. Agriculture Farm, Delakhadi, Chhindwara

191 Krashak Beej Uptadak Sah. Samiti, Chhindwara

192 Beej Utpadak Sah. Samiti, Thuniya, chhindwara

193 Sunrise Seeds, Imlikheda Chhindwara

194 M.P. State Seeds & Farm Development Corporation, Khamariya, Jabalpur

195 Bharat Krashak Samaj Samiti, Jabalpur

196 Balaji Seeds Gradar Unit, Suraj, Jabalpur

197 M.P. State Seeds & Farm Development Corporation, Dindauri, Mandla

198 M.P. State Seeds & Farm Development Corporation, Gadarwara, Narshinghpur

199 M.P. Oil Fed, Narsinghpur

200 M.P. State Seeds & Farm Development Corporation, Gadarwara, Narshinghpur

201 Shri Saai Beej Utpadak Sah. Samiti, Nandner, Narshinghpur

202 M.P. State Seeds & Farm Development Corporation, Seoni Malva, Hoshangabad

203 M.P. State Seeds & Farm Development Corporation, Chhapara, Seoni

204 M.P. State Seeds & Farm Development Corporation, Ashoknagar, Ashok Nagar

205 M.P. State Seeds & Farm Development Corporation, Bhind

206 Ms. Shankar Seeds Company, Bhind

207 M.P. State Seeds & Farm Development Corporation, Bhind

208 Govt. Agri Farm, Gohad, Bhind

209 M.P. State Seeds & Farm Development Corporation, Guna

210 M.P. State Seeds & Farm Development Corporation, Guna

211 Mohan Seeds P. Ltd., Madhu Sudangarh, Guna

212 Govt. Agri Farm, Mahukheda Gwalior

213 Chambal Ghati Hybrid Gwalior

214 Girraj Beej Udhyog, Gwalior

215 J.N.K.V.V., Morena

216 Chambal Ganga Seeds, Morena

217 M.P. Oil Fed, Kuthrawali, Morena

218 J.N.K.V.V., Morena

219 Sudhindra Company, Morena

220 Ambe Ji Beej bhandar, Morena

221 M.P. State Seeds & Farm Development Corporation, Baroda

222 S. Kumar Industries, Shivpuri

223 M.P. State Seeds & Farm Development Corporation, Shivpuri

224 Co-Ordinator, P.F.T.D.P.I., Pichhor, shivpuri

225 Shir Kishan Beej Utpadak Awam Prashankaran Sah. Samiti, Pohari Shivpuri

226 M.P. State Seeds & Farm Development Corporation, Chorhata, Rewa

227 Kisan Beej Utpadak Sah. Samiti, Godhar, Rewa

228 Govt. Agri Farm, Birhuliya, Rewa

229 M.P. State Seeds & Farm Development Corporation, Reora, Satna

230 Govt. Agri Farm, Gahavara, Satna

231 Dada Agro Private, Satna

232 Nahar Agro Seeds, Sohaval Satna, Satna

233 Rama Agro Beej Utpadak Sah. Samiti, Bandha, Satna

234 M.P. State Seeds & Farm Development Corporation, Narsharha Shehdol

235 Pushkar Agro, Shahdol

236 Govt. Agri Farm, Birhuliya, Shehdol

237 Govt. Agri Farm, Sidhi Sidhi

U.P. State Seeds Certification Agency, Lucknow

A Agriculture Department

1 Agra

2 Aligarh

3 Atarra, Banda

4 Sidhari, Azamgarh

5 Bagapar, Maharjaganj

6 Baharaich

7 Barabanki

8 Beribagh, Saharanpur

9 Basti Cantt, Basti

10 Belipar, Gorakhpur

11 Bilwa, Bareilly

12 Chandausi, Moradabad

13 Dhampur, bunor

14 Etawah,

15 Faizabad

16 Ghazipur

17 Gonda

18 Jamunabad, Lakhimpur Kheri

19 Rawatpur, Kanpur

20 Katiyar, Lucknow

21 Khairabad, Sitapur

22 Lodhirpur, Shahjahanour

23 Mauranipur, Jhansi

24 Rampur, Kushinagar

25 Orai, Jalaun

26 Rahimabad, Luchnow

27 Raibareilly

28 Sahjanwan, Gorakhpur

29 Sandilla Hardoi

30 Unnao

31 Chharaura, Mathura

32 Kokhraj, Allahabad

33 D.A.O., Varansi

34 RA.T.D.S., Hardoi

35 D.A.O., Fatehpur

36 D.A.O., Etah

37 D.A.O., Mainpuri

38 D.A.O., Bulandshahar

39 Rath, Hamirpur

40 Rampur

B Horticulture Department

1 Basticantt, Basti

2 Aliganj, Lucknow

C Animal Husbandary

1 Hastinapur, Meerut

2 Neel Gaon, Sitapur

3 Aata, Oraijalan,

4 Majhara, Lakhimpur-Kher

5 Babupgarh, Ghazibad

6 Aarazilines, Varanasi

7 Saidpur, Lalitpur

8 Nibelet Barabanki

9 Bharari, Jhansi

10 Aligarh, P.C.D.F.

D Agriculture University

1 N.D. Uni of Ag. & Tech. Kumarganj, Faizabad

2 C.S.A. Uni. of Ag. & Tech. Kalyanpur, Kanpur

3 S.B.P. Uni. of Ag. & Tech. Daurala, Meerut

4 Allahabad Agriculture Deemed Uni. Allahabad

E ICAR

1 Indian Institute of Pulses Research, Kanpur

F National Seeds Corporation Ltd.

1 Sikandra, Agra

2 Bareilly

3 Daurala, Meerut

4 Etawah

5 Nakah Jungle C.W.C. Campus, Gorakhpur

6 Dadangar, Kanpur

7 Sahjahanpur

8 Varanasi

G. U.P. Seeds Development Corporation Ltd.

1 Varanasi

2 Amethi, Sultanpur

3 Masaudha, Faizabad

4 Bilwa

5 Kalyanpur, Kanpur

6 Bhitatra, Jalaun

7 Kankankhera, Meerut

8 Aligarh

9 Sikandrarau, Aligarh

10 Gorakhpur

11 Dumariaganj, Sidharth Nagar

12 Bilaspur, Rampur

13 Sultanpurpatti, Rampur

14 Muzffarnagar

15 Kamlabad Badholi, Lucknow

16 Etawah

17 Pilibhit

H. State Farms Corporation of India

1 Girjapuri, Bahraich

2 Lalganj, Raibareilly

I Co-Operative

1 Iffeco Phulpur Allahabad

2 Kribhco Aligarh

3 Kribhco Moth, Jhansi

4 Kribhco Shahjahanpur

5 Kribhco Gazipur

6 Kribhco J.P. Nagar

J. Individual Seed Producers

1 Anil Kumar, Alipur, Meerut

2 Shivalik Seeds Sansthan Chilkana, Saharanpur

3 Associated Seed Growers Faizabad

4 Magardhaha Krishi Sansthan, Mirzapur

5 Malook Singh Seeds Farms, Rasoolpur, Ghazibad

6 Krishna Seeds Pvt. Ltd., Agra

7 Kartar Seeds Hasanwall, Saharanpur

8 Continental Seeds & Chemicals Varanasi

9 Etawah Beej Utpadan Kendra, Jaswantnagar, Etawah

10 P.L. Seeds, Jaswantnagar, Etawah

11 Panjab Seeds Farm, Kania, Ghazibad

12 Ramsaran, Seeds Farm, Kania, Ghazibad

13 Agarwal Seeds & Chemicals Rampur Road Bilaspur, Rampur

14 Chandan Seeds Farm, Chaubia, Etawah

15 Kaveyri Ganga Seeds, Varanasi

16 Oriental Seed Varanasi

17 Dwarika Seed Ramnagar, Varanasi

18 Pantnagar Tarai Agri Seeds Bilashpur, Rampur

19 National Agro Seeds Corporation India, Indira Nagar, Lucknow

20 Pragati Seeds Ganesh Ji, Jalaun

21 A.K. Seeds Ganesh Ji Jalaun

22 Subhash Seeds Corporation, Madhavganj Etah

23 M.P. Seeds Farm Mohanpur, Aetah

24 Narayan Seeds Dibiyapur, Etawah

25 Prasad Seeds Farm Rura Kanpur, Dehat

26 Bharat Beej Farm Bhandar Rampur, Manpur, Bulandshahar

27 B.K. Seeds India, Chhatari Bulandshahar

28 B.R. Seeds Company Mudgalpur Sarnath, Varanasi

29 Pilibhit Pant Nagar Beej Ltd., Bisalpuri Road, Pilibhit

30 Ankur Seeds & Agro Chem. Karonda Mohammadi Road, Shahjahapur

31 Dahia Seeds & Agricultural Farm Sawalpur Nawada, Saharanpur

32 S.S.P. Seedproduction Nindial Td. Jangapur, Pratapgarh

33 Chaudhary Seeds Farm Buxar, Ghazibad

34 Sonaganga Internationa Seeds Co. Jagatganj, Varanasi

35 Om Agra Tech Agrawal Bhawan, Achhnera, Agra

36 Kalu Seeds Farm Gram Dadayara, Ghazibad

37 Shivam Agro Tech Agrawal Bhawan, Achhnera, Agra

38 Prasad Seeds Corporation, Bareilly Morh, Shahjahanpur

39 Progressive Seeds Pvt. Ltd., 50 Tivarinath, Pilibhit Road, Bareilly

40 Nainital Agro Seeds Bilaspur

41 Lord Parshpati Seeds Pilibhit

42 Laxmi Seeds Jharkhand Nehtaur, Bijnor

43 Kalptarun Farm Pvt. Ltd., Umarpur, Jaunpur

44 A.B. Seeds & Agri. Chemical Atsaliya Road, Shahjanpur

45 Prabhat Hybrid Seeds Vankar, Varanasi

46 Swami Sri Seeds Chandra Chaumuhani Varanasi

47 Modern Tarai Seeds Pvt Ltd., Pilibbhit

48 Rajpoot Seeds Dakhinara Etah Road, Shikohabad

49 Ishwar Packaging Products Pvt. Ltd., Sultanpur

50 Shyam Seeds Etawah Road, Jashwanti Nagar, Etawah

51 Bajrang Seeds Farm, Auraia Road, Dibiyapur, Auraia

52 Shakti Agro Seeds, Rampur Road, Bilashpur, Rampur

53 Bhatia Quality Seeds, Jugurnagar, Baheri, Bareilly

54 Brijbhoomo Seeds Pvt. Ltd., Govardhan Road, Mahura

55 M.H. Tarai Seeds Milak Khanam Bilashpur Rampur

56 Ganesh Seeds Puranpur Road, Pilibhit

57 Vasheno Seeds Corporation Kunvar Garh, Pilibhit

58 Hadrabad Seeds Farm Chaukia Manaur, Mirzapur

59 U.P.Purvanchal Beej Industries Nakati Khuruhuja Chandauli

60 Samrat Beej Company Tadia Jakhani, Varansi

61 United Agro Seeds Corporation Indurpur Bilaspur, Rampur

62 U.P. Tarai Agro Seeds Bisharatpur Bilsapur, Rampur

63 bhawani Seeds & Biotechbhureka Mathura

64 Hariom Agro Industries Rampur, Mathura

65 Shakati Green Gems Pvt. Ltd., Narharia Basti

66 Thakur Prasad And Sons, Mirzapur

67 K.R.B.L. Ltd., Achhaja Gotambudh-nagar Basti

68 R.s. Seeds Farm Chaubia, Etawah

69 Krishna Kaveri Seeds Bhotapura, Chandauli

70 H.L.Seed Farm Jaswant Nagar, Etawah

71 Deepak Seed Farm Jaswant Nagar, Etawah

72 G.L. Seed Karhal Manpuri

73 Bajrang Tarai Seed, Pilibhit

74 Satyam Agro Seeds and Sar Chandauli

75 Mother Dropdi Seed Vidirka Aligarh

76 Om Seeds Company Sarai Dhanesh Gazipur

77 Pay Aur Line Seeds Sikanderabad Bulandshahar

78 Raj Sales Orai Jaluan

79 Nainital Tarai Seds Swar, Rampur

80 Pargati Enterprises Jamalganj, Swar Rampur

81 Greenline Seeds & Mohamadpur Rajoopi, Bijnor

82 Haldhar, Seeds & Farms Nagla Lodha Etah.

83 Hasanpur Sanyuot Sahkari Krishi Samiti Ltd. Gangsra Shahjahanpur

84 Kumar Seeds Puranpur, Pilibhit

85 Shree Shankar Tarai Seeds Maharajpur

86 shiwalic Agro Seed Sarwarangar Bilspur, Rampur

87 Shree Seeds 505/752, Alkapuri Etawah

88 Shree Seeds 505/752, Alkapuri Etawah

89 Om Ganga Kaveri Seeds Puranpur, Pilibhit

90 Narayan Agro Seeds Shikohabad Firozabad

91 Mangala Hybrid Seeds Bisauli, Pilibhit

92 Shir Krishna Seeds Puranpur, Pilibhit

93 Balaji Agro Seds, Mathura

94 Natural Indian Seeds Company Farrukhabad

95 Dharnidar Seeds Baswa Aligarh

96 Anu Seeds Nangla Padam, Aligarh

97 Kisan Seeds Corporation Bisauli Badaun

98 Harit Agro Seeds Corporation Bilaspur, Rampur

99 Guru Nanak Seeds Company, Bilaspur, Rampur

100 Panjab Seeds Bilaspur, Rampur

101 Hariwestern U.P. Seeds Bilaspur, Rampur

102 Kissan Tarai Seeds Bilaspur, Rampur

103 Keten Seeds Farms Karchhana Allahabad

104 Padam Seeds & Farms Allahabad

105 Ganga Seeds Coporation Bilaspur, Rampur

106 Hari Krishans Seeds Khirni Moradabad

107 Bhatia Seeds Jugroonagar Bareilly

108 Jai Bhole Seeds Bhabatpura Chandauli

109 Pantnagar Agro Seeds Products Thakurdwara Moradabad

110 Anupam Trai Seeds Corporation, Roopapurkripa Pilibhit

111 Agro Pantnagar Seeds bhedpur, Shahajahanpur

112 Kansal Trai Seeds, Bilaspur, Rampur

113 P.R.D.F. Khaililabad Santrabims Nagar

114 Bajrang Agro Seeds Math Khera Road, Bilaspur

115 Blue Diamond Industries Ltd., Begrapur, Muzaffarnagar

116 Shree Krisana Kaveri Seed Golagokarannath Kheri

117 R.K. Seeds Laharadeh Kapethi Varanasi

118 Haritama Seeds Jasvantnagar Etawah

119 Verma Interprises Arjunpur, Sitapur

120 Shribalaji Agro Seeds Kasimpur Power House, Aligarh

121 Haryana Agree Seeds Matkhera Road, Bilaspur, Rampur

122 M.P. Seeds Ghiror Road Jasarana, Firozabad

123 Pantnagar Seed Farm Chhapri Sikhkdi Ghazipur

124 N.H.R.D.F., Ishara Bhawan, Deoria

125 The Kumarn Seeds Farm Naqti Madihan, Mirzapur

126 Shanti Agretake Amaripur, Papaura Chandauli

127 Sharang Seeds Co., Nanhupur Saharnath, Varanasi

128 Goyal Seeds, 87, Mohrampur Mathura Road, Sichandra Agra

129 Dhara Seeds Farm Hardoiraja, Urgaun, Jalaun

130 Panday Seeds Pvt. Ltd., Faizabad

131 Sankatmochan Seed Company, Ghazipur

132 Genesh Agro Seeds Industries Dhanipur, Aligarh

133 Shrilaxmi Tarai Seeds Thakurdwara, Moradabad

134 Golden Seeds Products Bahari, Bareilly

135 Shiv Seeds & Farms Sidhauli Dhanipur, Aligarh

136 Shiv Ganesh Seeds, Jalaun

137 Rajdeep Seeds R.K. Puram A.D.A. Aligarh

138 Garg Tarai Agro Seeds, Bilaspur

139 Sri Balagi Sahara Seeds & Research Bilaspur

140 Annpurna Agro Seeds Bilaspur, Rampur

141 New Panjab Seeds & Processing Lant Bilaspur, Rampur

142 vijay Seeds & Farm Abharajpur Etah

143 Sri Krishna Radhey Agro Seeds Achhanera, Agro

144 Durga Tarai Seeds Math Khera Road Bilaspur

145 J.K. Agro Seeds Raza Nagar, Swar Rampur

146 Lata Mahila Upbhaukta Sahakari Samiti Ltd., Dhaneliuttari Milak Rampur

147 Tiffco Seeds Visanpur, Jageera Bilaspur

148 Ganga Gomati Seeds Corporation Hathapura, Bujuurg Sahjahanpur

149 Amrit Seeds Puranpur Bharokhurd Pilibhit

150 Ganga Vailly Green Seeds Dhanipur G.T. Road, Aligarh

151 Dev Seeds Gresiganj Chhibramau Kannau

152 Badashah Seeds Com, 103, Moh. Kheta Shikoha, Firozabad

153 S.D. Agro Industries Gsrhi Rambal Oil, Mathura

154 Aawasthi Seeds Corporation Pokharpur, Kanpur

155 Navkar Seeds Coporation Visharatpura, Bilaspur, Rampur

156 Kamakhya Kaveri Seeds Bhawatpura Baburi Chandauli

157 Greenland Seeds Corporation, Shivpuri, Varanasi

158 Amar Agro Seeds Shahpur Po. Ramraj Muzaffarnagar

159 Betwa Seeds Corporation Mo. Khandeyrao Jalaun

160 Shivshakti Agro Seds Yadav Nagar Bhood Bulandshar

161 Pateil Seds Chawanpura, Atc Jalaun

162 Agrawal Enterprises Hastinapur Road Ramraj, Meerut

163 Continental Seeds and Chemical Oral, Jalaun

164 Shivshakti Agro Seeds Yadav Nagar, bhood Bulandshar

165 Pateil Seeds Chawanpura, Atc Jalaun

166 Saubhari Seeds And Rarms Panethi Aligarh

167 Panatnnagar Uttan Seeds Products Devipur Gorakhpur

168 Punjab Hydrib Seeds G.P.O. Alipur Madra Ghazipur

169 Purvanchaliseed Company Gilaulidoob Kalan Sharawastibahr

170 Adarsh Andhra Seeds Corporation Bhawatpura, Baburi Chandali

171 Shree Ganesh Agro Seeds Kuyankhea Bilaspur, Rampur

172 Mahal Seed Farm Jarauthi Hapur, Ghazibad

173 Rajeev Seed Farm Dadayara Hapur Ghazibad

174 Ashok Seed Farm Shyampur Dadayara Hapur, Ghazibad

175 Surya Seeds Industries, Bisalpur Road, Pilibhit

176 Sri Gurunanad Seeds Bhusainga Baheri Bareilly

177 Sona Seeds Agro Products Rampur Road, Nagina, Bijnor

178 Raj Agro Seeds Farm, Nagla Kalu, Mathura

179 Nainital Agro Seeds Chumuha, Mathura

180 Shiva Seeds Company Sihora Mathura

181 Sharda Seeds Puranpur, Pilibhit

182 Kishan Seeds Corporation Bisauli Badaun

183 Sri Sai Tera Seeds Company Dhanaur Bilaspur

184 gurudeo Seed Production Centre Satharia Jaunpur

185 Swami Seeds Farm Kewala Raja Kabagh Etawah

186 Haryana Seeds Company, Mihipurwa, Bahraich

187 Durga Mini Rice & Seed Processing Centre, Krishna

188 India Seed Farm Pandeypur, Varanasi

189 Sood Seed, F-57, Ram Nagar, Chandauli

190 Bundelkhand Seed Moh-Katra Jalaun

191 Vindhiyachal Seeds Bikapur Sikandarpur - Chandauli

192 Paras Seeds Corporation, Loanch, Jalaun

193 Triveni Seeds Mirazapur

194 Agri Clinic Agri Business Ranepur Chandoli

195 Jindal Seed Hapur Ghaziabad

196 New Poineer Seeds, Bulandshahar
197 Narmada Seeds Jaswant Nagar, Etawa
198 R.K. Seeds Jaswanti Nagar, Etawa
199 Amit Seed Pachipura Jalaun
200 Asha Seeds Moth Dev Nagar, Jalaun
201 Samay Agro Pvt. Ltd., Sidharth Nagar
202 Kuber Vegitable, Jalaun
203 Aditya Seeds Udyog Ind. Konch Jalaun
204 U.P. Agro Hybrid Seeds Sikoha Bad Firojabad
205 B.L. Seeds, Jalaun
206 Mudit Seeds, Aligarh
207 Ansul Agro Seeds, Kasganjeta
208 A.J.S. Seeds, Shahjanpur
209 Dhan Versha Agro Seeds, Bilaspur
210 Shubham Seeds, Mathura
211 Maharana Seeds, Mathura
212 Neelam Seeds Baitra, Jalaun
213 Annapoora Rice Mill Thakur Dwara, Moradabad
214 Sudha Seeds, Deoria

Maharashtra State Seed Certification Agency, Akola

1 Seed Processing Plant MSSC, Shioni, Akola
2 SPP & Ginning Centre MSSC Shioni, Akola
3 Seed Processing Plant Vegetable, Shioni, Akola
4 HOD Dr. PDKV Agriculture, Eng. SPP, Akola
5 Chief Seed Production Officer, Dr. PDKV (SPP) Wani Rambhapur, Akola
6 Chief Seed Production Officer, Dr. PDKV (Ginning Center) Wani Rambhapur, Akola
7 Ginning Center, MSSC, Patur, Akola
8 The New Akot Cotton Ginning And Pressing Co. Ltd., Akot, Akola
9 SPP, MSSC, Washim
10 SPP Mahalaxmi Seed Co. M.A.I.D.C., Akola
11 SPP Anup Seed Co. Kurum, Akola
12 Basant Seed Kanheri, Akola
13 Mahesh Hy. Seed Co., Akola
14 Nuzividu Seeds Co. Akola
15 Tirupati Seeds Co. Malegaon, Washim
16 Anupam Agro Industries M.A.I.DC., Akola
17 Poonam Seeds Co., Washim
18 Shrikant Cotton Industries G.I.P. Unit Shioda, Akola
19 Balaji Co-Op Sugar Ind. Masalapen, Washim
20 Gauri Seeds & Bio Tech Corporation Chikhalgaon, Akola
21 Janki Seeds & Research Pvt. Ltd., Akola
22 Nishant Sahakari Gramin Patpurawa-rtha Sanstha, Chohotta Bajar, Akola
23 SPP, MSSC, Tapowan, Amaraoti
24 National Seeds Corporation, NSC, Amaraoti
25 Gramin Krishi SPP Asegaon Purna, Amaraoti
26 Amol Seeds Co. Chandikapur, Amaraoti
27 Shiv Agro, Waknath, Tq, Dhamangaon, Amaraoti
28 Shree Seeds Co. Amaraoti
29 Maharashtra Agro Seeds Bhatkuli Amaraoti
30 Mahalaxmi Seeds Co. Masod, Amaraoti
31 Hindustan Agro Servises SPP Badnera, Amaraoti
32 Ravi Seeds Co., Amaraoti
33 Daryapur Sahakari Gining and Pressing Factory (MSSC) Amaraoti
34 Vidharbh Agro Seeds Co. Walgaon, Amaraoti
35 The Vidharbha Co. Op So. Ltd., Maphed MSSC, Amaraoti
36 JDS Ginning Sategaon Tq Anjangaon, MSSC, Amaraoti

37 Malani Ginning & SPP, Yeotmal
38 Libarti Seeds Co. Yeotmal
39 SPP Butibori, Nagpur
40 Ankur Seeds Co., Nagpur
41 SPP MSSC Shelu, Wardha
42 Yeshoda Hy. Seeds Co., Kumbhi, Wardha
43 SPP MSSC Mul, Chandrapur
44 SPP, MSSC Gadeaon, Bhandara
45 Daftari Agro Pvt. Ltd., Shelu, Wardha
46 Dwarka Seeds, Bajargaon, Nagpur
47 Nuzevedu, Seeds Pvt. Ltd. Gadchiroli
48 Purti Agro Tech SPP Dhapewada, Nagpur
49 Arun Hy. Seeds Pvt. LTd. Rotha, Wardha
50 SPP, MSSC, Chikhali, Buldhana
51 SPP, MSSC, Malkapur, Buldhana
51 SPP, MSSC, Khamgaon, Buldhana
52 Ankur Seeds Pvt. Ltd., Malkapur, Buldhana
53 Mahyco Seeds Ltd., Dhanora, Buldhana
54 Golden Seeds Co. Chikhali, Buldhana
55 Akshya Seeds SPP Malgani Tq. Chikhali Buldhana
56 Balaji SPP, Shindi Harali, Tq. Chikhali, Buldhana
57 Harit Kranti Seeds Preccessor, Deolgaon Raja, Buldhana
58 Sheetal Seeds, SPP, Deolgaon Raja, Buldhana
59 Birla Agro PVt. Ltd., Khamgaon, MSSC, Buldhana
60 Buldhana Urban Seed, Chandaknagar, Dongarkhandal, buldhana
61 SPP MSSC, Parbhani
62 The Sahayog Shakari Kharadi Vikri Sangh Ltd., Parbhani
63 Mararthwada Krishi Vidhyapeeth SPP, Parbhani
64 Balaji Seeds Co. Parbhani
65 Balaji Hy. Seeds Corporation, Nanded
66 Ganesh Seeds Company, Nanded
67 SPP MSSC, Cotton, Dhanegaon, Nanded
68 Maruti Seeds Company, Nanded
69 Parsewar Seeds Co. Nanded
70 Samarth Seeds Pvt. Limited, Nanded
71 Laxmi Hybrid Seeds Co. Ltd., Nanded
72 Santosh Seeds Company, Nanded
73 SPP MSSC Dhoki, Osmanabad
74 SPP DMO Godwon, Parbhani
75 Kharedi Vikri Sangh Godwon Hingoli
76 SPP Texcom MIDC, Nanded
77 Anil Agro Agencies, Jawalabajar, Hingoli
78 SPP MSSC, Erandol, Jalgaon
79 Ginning Center, Pachora, Jalgaon
80 Arihant Ginning And Pressing Factory Dongar Kathora, Jalgaon
81 National Seeds Corporation, Jalgaon
82 Bhagashri Agro Inputs, Jalgaon
83 Kohinoor Seeds Pvt. Ltd., Pachora, Jalgaon
84 SPP Agriculture College, Dhule
85 SPP MSSC, Satana, Nashik
86 National Horticulture Research & Development, Nashik
87 Loknayak Jayprakash Narayan Sahkari Soot Girni, Untawad, Nandurbar
88 Shri Salasar Cotton Industries, Pachora, Jalgaon
89 Sahayog SPP, Tandulwadi Tq, Rawer, Jalgaon
90 SPP Seed Multiplication Farm, Musalgaon, Nashik
91 National Seed Corporation, Aurangabad
92 Green Gold Seeds Ltd, Aurangabad
93 Ajit Seeds Pvt. Ltd., Chitegaon, Aurangabad
94 Krishidhan Seeds Jalna-1
95 Krishidhan Seeds Pvt. Ltd., Jalna-2
96 SPP MSSC, Jalna
97 Anand Ginning & Preesing Ind. Pathrod, Beed

98 Mulchand Fulchand Krishi Uyog Ltd., Jalna
99 Jaybhavani Seeds And Research Ltd Partur, Jalna
100 Jindal Traders, SPP, Jalna
101 SPP Khandala , Shrirampur, A. Nagar
102 SPP Market Yard Shrirampur, A. Nagar
103 Mahatma Fule Krishi Vidhyapeeth, Rahuri, A. Nagar
104 Mauli Agro Servises, Kopergaon, A. Nagar
105 Superintendent, Agriculture College, Manjri, Pune
106 Solapur District Milk Federation, Tembhurni, Sholapur
107 Shivamrut Milk Federation, Akluj, Sholapur
108 National Seeds Corporation, Sangli
109 SPP MSSC, Ashta, Sangli
110 Sheti Padwidhar Sahakari Sangh Ltd,. Kolhapur
111 Dr. Babasheb Sawat Kokan Krishi Vidhyapeeth, SPP Wakawali, Dapoli
112 Deepak Agro Solutions Ltd. Wagholi, Pune
113 Hy. Seeds Co. Dudhediya, Shrirampur, A. Nagar
114 Shiva Seeds, Rahuri, A. Nagar

Chief Seed Certification Officer, Seed Certification Agency Govt. of NCT Khaiberpan, Civil Lines New Delhi-54

01 G-Tech Agro Industry Alipur Delhi
02 Farmer Seed Consumer Federation R Z – 14 Nagabyath New Delhi

Assam State Seed Certification Agencies

01 B N College of Agriculture
02 RARS N Lakhimpur
03 CPRS Upperb shilong Meghalaya
04 KVK Napam Tejpur
05 State Seed Farm Sambrak Deptt of Agriculture Garo Hills Meghalaya
06 Green Tech Agro Industry Ambagan Nagaon
07 RARS Shillongoni
08 High Tech Green Harvest Guwathi
09 RARS Akberpur Cachar
10 KVK Cachar Arunachal
11 Duralavcharn Kurimgang
12 Mahakal Field Trial Station Mahakal Karimgang
13 Moneirkhal Seed Farm Sachinpur Cachar
14 Field Trial Station Khetri
15 Expertimental Research Station Meghalaya

Rajasthan state Seed and Organic Produce Certification Agency Jaipur

Plant No.	Unit Name	P Agency	Address	Contact No.
1	Jaipur	R.S.S.C., Durgapura, Jaipur	ARS Campus, Durgapura, Jaipur.	0141-2550614
3	Ajmer	RSSC-Ajmer	Tabiji Farm, Bewar Rd., Ajmer	0145-2440830

4	Chittorgarh	RSSC-Chittorgarh	Ochhadi, Nimbaheda Rd., Chittorgarh	01472-241012
5	Chittorgarh	RSSC-Udaipur	New Mandi Yard, Udaipur	0294-2584687
6	Jodhpur	RSSC-Jodhpur	Opp. Mandore Railway Station, Jodhpur	2570692
7	Jodhpur	RSSC- Sumerpur	RIICO Ind. Area, Sumerpur	02933-258326
10	Kota	MAF, Ummed Ganj	MPUAT, Ummedganj Farm, Kota.	0744-2209388
11	Alwar	RSSC	Naya Bas, Alwar	0144-2701913
14	Bharatpur	NSC-Bharatpur	E12 &13, RIICO Ind. Area, Bharatpur	05644-222713
15	Sriganganagar	R.S.S.C., SGNR	Karni Marg, SGNR.	2440070
16	Sriganganagar	Janta Sgri Seed	Opp. Union Motors, Shakti Marg, SGNR.	2460010
17	Suratgarh	RSSC	Ind. Area, Suratgarh	01509-220206
18	Suratgarh	CSF-Suratgarh	CSF, Suratgarh (Block-II)	
19	Suratgarh	CSF-Suratgarh	CSF, Suratgarh (Block-I, Rang Mahal)	
20	Suratgarh	CSF-Sardargarh	CSF, Sardargarh.	01509-283027
21	Suratgarh	CSF-Jetsar	CSF, Jetsar.	01498-261233
22	Sriganganagar	Bihani seeds	2ML, Nathawali, SGRN,	2464839
23	Kherli	Krishna Agro Service Center	Near Railway Crossing, Kherli.	01492-220381
24	Kota	R.S.S.C., Kota	Karkhana Bagh, Near Police Line, Kota	0744-2326736, 0744-2328133-P
25	Bhilwara	RSSC-Bhilwara	Near - Arjiya Farm, Bhilwara.	01482-264684
26	Kota	N.S.C.-Kota	E-134, Indraprasth Ind. Area, Rd.2, Kota	0744-2425954
27	Sriganganagar	N.S.C.	Ind. Area, Abohar Road., SGNR.	2440232
28	Chittorgarh	NSC-Chittorgarh	Cotton Ginning Factory, Bhilwara Road, Chittorgarh	01472-245087
30	Ajmer	NSC-Ajmer	F45-46, Prabhatpura Industrial Area, Ajmer	0145-2695294, 2620040
33	Sriganganagar	Singhal Seed Co.	Near Tiny Tots School, Behind Civil Hospital, SGNR	
35	Kota	Goyal seeds	B-43, Indraprasth Ind. Area, Rd.No.2, Kota	0744-3291737

36	Bharatpur	Rajasthan Agricultures	F-59, RIICO Ind. Area, Bharatpur	9414370814
37	Suratgarh	CSF-Sardargarh	CSF, Sardargarh.	01498-264773
44	Sriganganagar	National Seeds	HN-15, Suratgarh Road, Sriganganagar	
47	Sriganganagar	Sriram Seeds	Motor Market, Suratgarh Road, SGRN.	2442438
48	Pilibangan	Kuber seeds	P.No.10-11, Ind. Area, Pilibanga.	01508-233260
49	Sriganganagar	Vishnu seeds	121, Railway Crossing, Karni Marg, SGNR.	2440805
51	Sriganganagar	M.R. Seeds Pvt. Ltd.	P.No.40 to 42, Ind. Area, Agarsan Nagar, SGNR.	9352084793
52	Bharatpur	Shiv Seeds	79, RIICO Ind. Area, Bharatpur	9414715129
53	Pilibangan	Northern Seeds	4-5, Ind. Area, Pilibangan.	01508-233343
54	Kota	Tilam Sangh	Near Naya Goan, Rawat Bhata Road, Kota	0744-2470083
56	Pilibangan	Vinayak Seeds	P.No. 42 to 44, Ind. Area, Pilibanga.	01508-233053
56	Jaipur	Maruti seeds	G-I, 215, Sitapura Ind. Area, Jaipur.	935680685
57	Hanumangarh	Mittal Seeds & Chemicals Pvt. Ltd.	C-12, Ind. Area, Hanumangar Junction.	01552-514294
59	Pilibangan	Oswal Seeds	63, Ind. Area, Pilibanga.	9414094635
60	Sriganganagar	Ahoka Seeds & Gen. Ind., Padampur	33 to 36, Ind. Area. Padampur	
62	Sriganganagar	Deol seeds	Chak-7-Z, Near Old Choongi, Karanpur Road.	9314677870
66	Sriganganagar	Maharaja Seeds	Motor Market, Suratgarh Road, SGRN, Ph.No.2445971	
67	Sriganganagar	Sangam Seeds	98, Ind. Area, Agarsan Ng., SGNR.	9413713597
68	Sriganganagar	Laxmi Seeds Corporation.	92 to 94 Ind. Area, Agarsan Ng., SGRN	
71	Sriganganagar	R.K. seeds	Opp. Union Motors, SGNR.	2460653
74	Banswara	R.S.S.C.	Near Sant Paul School, Tikriya, Banswara	243210
77	Suratgarh	RSSF & P-Suratgarh	Near Bhagwan Sar.	
80	Kherli	Shreeji Udyog	H 20-21, Ind. Area, Kherli.	01492-221451

83	Sriganganagar	Jai Shanker Seeds	P.No. 9, Industrial Area, Abohar Road, SGNR.	9314441944
87	Gharsana	Swastic seeds	Ind. Area, Kupali Road, Gharsana.	01506-221711
88	Pilibangan	Prem Cotton Ginning, Goluwala	Padampur Road, Goluwala.	01508-220024
91	Sriganganagar	H.S. Seeds	6-C, Motor Market, SGNR.	941409052
92	Kota	KRIBHCO	19-B, Indraprasth Ind. Area, Rd.2, Kota	0744-2433010
94	Gharsana	RCDF-Bikaner	RCDF Seed Processing Unit, Rampura Basti, Lalgarh, Bikaner.	0151-2251738
95	Alwar	Surya Seeds	Village-Ballabora, Vijay Mandir Road, Alwar	0144-3208555
96	Alwar	Unnat Agro Input	Shikari Ka Bas, Vijay Mandi Road, Alwar	2730185
97	Banswara	Haldi Ghati Seeds	P. No. 61-62, Ind. Area, Tikriya, Road-1, Opp. Police Line Banswara	
99	Kota	Amit Industries	F-559, Indraprasth Ind. Area, Rd.6, Kota	0744-2490735
100	Jodhpur	Anupam seeds	Gehlot Krishi Farm, Near Police Station, Nagaur Bye Pass Road, Jodhpur	2577360
101	Bharatpur	Uttam Agritech.	Gangapur	07463-231911
102	Kherli	Chandra Agro Seeds	H22-23, Ind. Area, Kherli.	01492-220593
103	Gharsana	Deepak enterprises	G-9, RIICO Ind. Area, Gharsana.	01506-220058
105	Sriganganagar	Adhunik Udyog, Padampur	27 to 32, Ind. Area, Padampur	
107	Sriganganagar	Singhal Seed Grower	Opp. Union Motors, SGNR.	2460694
108	Bharatpur	RSSC-Bharatpur	G-44 & 45, RIICO Ind. Area, Bharatpur	9413385320, 05644-223792
109	Sriganganagar	Shri Ganpati Seeds	Near Suratgarh Bye-pass, SGNR.	2443685
110	Gharsana	Shri Ballabh Seeds	7A, RIICO Ind. Area, Gharsana.	01506-221247
112	Jaipur	Shubh Sree Seeds	B-187, Rd. No.9, VKIA, Jaipur.	3268564
113	Ajmer	Greenvelly Seeds-	Bhilwara	

114	Bharatpur	Azad seeds	Old Ind. Area, Jaghina Road, Bharatpur	9414208488
116	Jodhpur	Bhagirath agro trading	F-20, Ind. Area, Mandore, Jodhpur	9928905455
117	Jaipur	Rajdhani Seeds-Renwal	E8-10, RIICO Ind. Area, Renwal	01424-221359, 9414518513
118	Sriganganagar	National Seeds	G-I/143C, Udyog Vihar, RIICO, Sriganganagar	
119	Sriganganagar	Shakti Agrotech (India)	7-E, Hanumangarh Road, SGRN.	
121	Kota	Navin Industries	B-39, Small Scale Industries Area, Kota	0744-2362398
123	Hanumangarh	RSSC-Hanumangarh	Rawatsar Road, Near ATC Farm, Hanumangarh Town.	01552-230407
124	Kota	R.S.S.C., Baran	Infront of College, Grain Mandi Yard, Kota	07453-230632
125	Kherli	Baldev Seeds	Village-Arruva, Tehsil-Katumar, District-Alwar.	9414499697
126	Jaipur	Shri Balaji Agri.	F-116, Ind. Area, Sikar	
127	Pilibangan	Maheshwari Fertilizer		
128	Bharatpur	Baba Seeds		9352637801
129	Jodhpur	Krishna Seeds		
130	Sriganganagar	Gopal Agro	78, Ind. Area, Agarsen Ng. SGNR.	2460126
131	Bharatpur	Shubh Sankalp,	Sapotra	
132	Sriganganagar	Kewal Krishan Rahulakumar	78-Old Dhan Mandi, Raisingh Nagar.	01507-220629, 9414091949
134	Sriganganagar	M.K. Seed, Karanpur	Gaj Singh Pur Road, Near Railway Crossing, Sri Karanpur.	
135	Kota	G.R.S. Biotech	19-B, Indraprasth Ind. Area, Rd.2, Kota	9414055958
136	Kherli	Shailja enterprises	Rampura Rd. Kherli.	01492-222438
137	Sriganganagar	Gayatri Seeds		
138	Hanumangarh	Waheguru Seeds	D-17, Ind. Area, Phase-I, Hanumangarh Junction.	01552-260481
139	Sriganganagar	Jai Shanker Seeds	P.No.3C & 6C, Industrial Area, Abohar Road, SGNR.	9314441944
140	Alwar	Aarson Agrotech	Govindgarh Mod, Delhi Rd., Ramgarh, Distt.-Alwar	01468-232526

141	Hanumangarh	Ganesh & Co.-Sangaria	36 to 37, RIICO-IInd Phase, Sangariya.	9214742174
142	Pilibangan	Jaishree Seeds	P.No.50, 52 & 54, Ind. Area, Pilibanga.	9414578311
143	Sriganganagar	Krishak Hybrid Seeds	Opp. Tata Motors, Shakti Marg, SGNR.	9413072720
144	Pilibangan	Arham Seeds	38 to 41, Ind. Area, Pilibanga.	01508-235697
146	Suratgarh	Shri Ganesham Seed	34, A-Block, Opp. Jhulalal Mandir, Jetsar.	01498-264216
147	Sriganganagar	Kisan Seeds	6, Ind. Area, Agarsan Nagar, SGNR.	
148	Jaipur	Sarswati Hybrid	NH-8, Balaji Mod, Bagawas Ahiran-303119, Shahpura.	
149	Sriganganagar	Bharat Seeds	P.No.12, Ind. Area, Agarsen Nagar, SGNR.	2484640
150	Banswara	Satguru Seeds	Village-Shivpura Doongarpur Road, PO-Kupadiya, Distt-Banswara	
151	Gharsana	Kishanlal Shyam Sunder	F-126, Karni Ind. Area, Bikaner.	0151-3209943
152	Kota	R.S.S.C., Jhalawar	Near Khel Sankul, Khanpur Rd., Jhalawar	07432-211205
153	Kota	Balaji Seeds	Silore, District-Bundi	9829036106
154	Sriganganagar	Raghav Seeds	Ind. Area, Agarsan Nagar, SGNR	
155	Sriganganagar	Sri Balaji Seeds	45, Ind. Area, Agarsan Ng. SGNR	
157	Jaipur	RSSC-Duni	Tehsil-Deoli, District-Tonk	

Details of Seed Processing Units in Tamilnadu

S.N.	S.P. Unit N.	Address	District
1	14	SSF, Pongalur Tirupur	Coimbatore
2	51	SSF, Pappankulam, Udumalpet	Coimbatore
3	680	AEC Periyanaickan Palayam, Coimbatore (North)	Coimbatore
4	682	AEC Tondamuthur, Coimbatore (South)	Coimbatore
5	688	AEC Annur, Avanashi Taluk	Coimbatore
6	689	AEC Avanshi, Coimbatore	Coimbatore
7	692	AEC Mettupalayam	Coimbatore

S.N.	S.P. Unit N.	Address	District
8	707	AEC Kinathukadavu	Coimbatore
9	738	AEC Madukkari	Coimbatore
10	742	AEC Pollachi (North)	Coimbatore
11	757	AEC Anamalai, Pollachi	Coimbatore
12	795	AEC Sulur, Palladam	Coimbatore
13	834	AEC Pollachi (South)	Coimbatore
14	870	AEC Nachipalayam	Coimbatore
15	879	AEC Perur	Coimbatore
16	927	AEC Pongalur	Coimbatore
17	936	AEC S.S. Kulam	Coimbatore
18	950	Central Godown, Coimbatore	Coimbatore
19	990	AEC Udumalpet	Coimbatore
20	1011	AEC Madathukulam, Udumalpet	Coimbatore
21	1018	AEC Sultanpet Palladam	Coimbatore
22	1019	AEC Palladm	Coimbatore
23	1021	AEC Tirupur	Coimbatore
24	1035	AEC Pedappambatti	Coimbatore
25	1038	AEC Alandurai	Coimbatore
26	1117	AEC Perumanallur	Coimbatore
27	1192	AEC Karamadaia	Coimbatore
28	1451	AEC Kumaralingam, Udumalpet	Coimbatore
29	87	Tucas - Thudiayalur	Coimbatore
30	408	National Seeds Corporation, Kurichi	Coimbatore
31	1142	Kribhco chennai, P.C.M.S. Pollachi	Coimbatore
32	177	Suja Seeds Corp, Avanashi	Coimbatore
33	192	V.Gopalsamy, Thekkupalayam	Coimbatore
34	208	Kovai Kadir Seeds, Arasur,	Coimbatore
35	234	Sri Santhalakshmi Mills Pvt Ltd. Pollachi	Coimbatore
36	334	SIMA, Udumalpet, Coimbatore	Coimbatore
37	378	The Consolidated Seeds company, Sanganur	Coimbatore
38	409	Thirumurugan Agro Seeds, S.S.Kulam	Coimbatore
39	433	Sakthi Sugars, Pollachi	Coimbatore
40	451	Sri Murugan Hybrid Seeds, Avinashi	Coimbatore
41	457	Sri Palani Andavan Ginning Factory, Sulur, CBE	Coimbatore
42	506	Adhi Seeds, Arasur, Coimbatore	Coimbatore
43	524	Siva Seeds, Iyer thottam, Palladam,	Coimbatore

S.N.	S.P. Unit N.	Address	District
44	537	Texcity hybrid seeds, Theethi palayam,	Coimbatore
45	538	Sri Sun Seeds, Avinashi	Coimbatore
46	539	Sri Krishnavilas Ginning Factory, Palladam	Coimbatore
47	558	Ganga Seeds, Kanur Pudur, Avinashi,	Coimbatore
48	583	Jai Seeds, Srinivaspuram, Alandurai,	Coimbatore
49	584	Sri Vishnu Enterprises, Peelamedu,	Coimbatore
50	590	P.V. Venkatesan, Mudalipalayam	Coimbatore
51	602	Vasantha Seeds, Kanur	Coimbatore
52	636	Sri Vignesh Ginning Factory, Annur	Coimbatore
53	639	Cheran Seeds, Kattambatty,	Coimbatore
54	1060	Sri Kumarn Hybrid Seeds, Avinashi	Coimbatore
55	1133	Saravana Hybrid Seeds, Kottamangalam	Coimbatore
56	1167	Kadir Agri Products, M.K. Pudur	Coimbatore
57	1175	Sri Vari Ginners, Avinashi	Coimbatore
58	1225	Sri Ramkumar prasanna Ginning Factory, Pollachi	Coimbatore
59	1243	Sri Srinivasa Ginning Factory, Annur	Coimbatore
60	1302	Ganesh High Tech Seeds, Udumalpet	Coimbatore
61	1303	Gangai Farm Seeds, Udumalpet	Coimbatore
62	1313	Sri Sabari Seeds, Avinashi	Coimbatore
63	1409	Krishna Ginning Factory, Annur	Coimbatore
64	1411	Sri Ramakrishna Ginning Rice Mill, Pongalur,	Coimbatore
65	1415	Bloom Bio Tech	Coimbatore
66	1417	Nandhi Seeds, Pappanaickanpalayam	Coimbatore
67	1436	Nachiar Corporation, Jameen uthukuli	Coimbatore
68	1447	Agro Seeds Corporation, Avinashi	Coimbatore
69	736	AEC, Panjayath Union Compound, Erode	Erode
70	731	ADO Erode, Sub AEC, Chittode	Erode
71	992	ADO Modakkurchi. AEC Modakkurchi	Erode
72	1039	ADO Modakkurchi. AEC Arachalur	Erode
73	717	ADO Kodumudi, AEC Kodumudi	Erode
74	743	ADO Kodumudi, AEC Sivagiri	Erode
75	758	ADO Perundurai, AEC Perundurai	Erode
76	716	ADO Perundurai, AEC Pethampalayam	Erode
77	968	ADO Chennimalai, AEC Chennimalai	Erode
78	691	ADO Uthukkuli, AEC Uthukkuli	Erode
79	1022	ADO Bhavani, AEC Bhavani	Erode

S.N.	S.P. Unit N.	Address	District
80	993	ADO Bhavani, AEC Kavundapadi	Erode
81	984	ADO Ammapettai, AEC Ammapettai	Erode
82	734	ADO Ammapettai, AEC Chanichandai	Erode
83	732	ADO Ammapettai, AEC Olagadam	Erode
84	733	ADO Anthiyur, AEC Anthiyur	Erode
85	735	ADO Anthiyur, AEC Athani	Erode
86	764	ADO Gobi, AEC Gobi	Erode
87	763	ADO Gobi, AEC Kugalur	Erode
88	761	ADO Gobi, AEC Vellankovil	Erode
89	875	ADO Nambiyur, AEC Nambiyur	Erode
90	894	ADO Nambiyur, AEC Kurumathur	Erode
91	766	ADO T.N.Palayam, AEC T.N.Palayam	Erode
92	760	ADO T.N.Palayam, AEC Kallippatti	Erode
93	808	ADO T.N.Palayam, AEC Nalroadu	Erode
94	766	ADO T.N.Palayam, AEC Kasipalayam	Erode
95	693	ADO Sathy, AEC Sathy	Erode
96	985	ADO Thalavadi, AEC Thalavadi	Erode
97	706	ADO Bhavanisagar, AEC Bhavanisagar	Erode
98	852	ADO Kangeyam, AEC Kangeyam	Erode
99	856	ADO Vellakovil, AEC Vellakovil	Erode
100	741	ADO Dharapuram, AEC Dharapuram	Erode
101	739	ADO Mulanur, AEC Mulanur	Erode
102	740	ADO Kundadam, AEC Kundadam	Erode
103	1027	Agirculture Officer, S.O.S.F. Bhavanisagar	Erode
104	313	Agriculture Officer, S.S.F, Sathy	Erode
105	114	Agriculture Officer, S.S.F, Bhavani	Erode
106	134	Secretary, G.Agri Co-op. M.So. Gobi	Erode
107	179	Secretary, Dharapuram Agri-Pro-Cop-M So	Erode
108	522	Sri Venkateswara Seeds, 63, Ellis Nagar, Dharapuram	Erode
109	441	Senthil Seeds, 73, Kolathuppalayam, Dharapuram	Erode
110	640	Praveen Seeds, 73 Kolathuppalayam, Dharapuram	Erode
111	596	Sakthi Seeds, 316, Kondaraspalayam, Dharapuram	Erode
112	983	Sakthi Seeds, 316, Kondaraspalayam, Dharapuram	Erode
113	1037	Sri Durga Seeds, 3/135, Varappalayam, Dharapuram.	Erode
114	1414	Sri Durga Seeds, 102-03 Bypass Road, Ellis Nagar, Dharapuram	Erode

S.N.	S.P. Unit N.	Address	District
115	1467	Nandhini Seeds, 69, Kariayur, Dharapuram	Erode
116	1036	Selva Seeds, 107/ 2A-Karur Road, Dharapuram	Erode
117	1298	Sri Devi Seeds, 153, Achiyur, Dharapuram	Erode
118	1468	Arun Ginning Mills, 215, Kolinjivadi, Dharapuram	Erode
119	1247	Amaravathi Seeds, 3/81 Nanjiyampalaym, Dharapuram	Erode
120	1412	Saravana Seeds, Gandhiji Nagar, D- Kalipalayam, Dharapuram	Erode
121	1483	Surya Seeds, Main Road, Nanijampalayam, Dharapuram	Erode
122	1509	Sri Krishna Seeds, 297- Udumalai Road, Dharapuram	Erode
123	614	Universal Seeds, Mettupalayam (P.O), Vellakovil, Dharapuram	Erode
124	1304	Sabari Seeds, 3/457, Modachur (P.O), Gobichettipalayam	Erode
125	1245	Ram Seeds, 13-B K.P. Extension, Gobichettipalayam	Erode
126	611	Gobi Seeds, O.N.Palayam, Odathurai, Gobichettipalayam	Erode
127	1174	Rani Seeds. Rangasamudram, Sathy	Erode
128	554	Gemini Hybrid Seeds, 7/338, Nehru Nagar, P.Puliampatti	Erode
129	331	Agro Seeds Corporation, 221, Main Road, P.Puliampatti.	Erode
130	597	Sri Krishna Seeds, 258, SPK Thottam, Sathy	Erode
131	417	Priya Seeds, Varathampalayam, Sathy	Erode
132	605	K.S. Seeds, 2/78 Mathampalayam Road, P.Puliampatti	Erode
133	422	Vijay Seeds, Sellappampalayam, P.Puliampatti	Erode
134	1166	Vijay Seeds, Sellappampalayam, P.Puliampatti	Erode
135	479	Sri Rangavilas Ginning Factory, 360, Sathy Main Rd, Puliampatti	Erode
136	1518	Sri Amman Seeds, 56-Palani Road, Kolinjavadi, Dharapuram	Erode
137	1522	Sri Kumarn Seeds, 1/146, Kovai Main Rd, N. Palayam (P.O) Dharapuram	Erode
138	824	ADA Salem, AEC Salem	Salem
139	881	ADA Salem, AEC Yercaud	Salem
140	878	ADA Salem, AEC Veerapandy	Salem
141	798	ADA Salem, AEC Panamarattupatty	Salem
142	926	ADA Salem, AEC Ayothiyappattanam	Salem

S.N.	S.P. Unit N.	Address	District
143	995	ADA Salem, AEC Valappady	Salem
144	36	ADA Omalur, SSF Danishpet	Salem
145	886	ADA Omalur, AEC Omalur	Salem
146	1122	ADA Omalur, AEC Karuppur	Salem
147	1120	ADA Omalur, AEC Kadayampatty	Salem
148	1094	ADA Omalur, AEC Tharamangalam	Salem
149	1125	ADA Mettur, AEC Kolathur	Salem
150	1104	ADA Mettur, AEC Nangavalli	Salem
151	883	ADA Mettur, AEC Mecheri	Salem
152	887	ADA Mettur, AEC Vellar	Salem
153	243	ADA Mettur Dam, SPU Mettur Dam	Salem
154	1316	ADA Mettur Dam, District Central Godown Mettur	Salem
155	1033	ADA Sankari, AEC Sankari	Salem
156	884	ADA Sankari, AEC Thevur	Salem
157	1013	ADA Sankari, AEC Idappadi	Salem
158	1095	ADA Sankari, AEC Poolampatty,	Salem
159	981	ADA Sankari, AEC Konganapuram	Salem
160	933	ADA Sankari, AEC Magudanchavadi	Salem
161	829	ADA Attur, AEC Attur	Salem
162	1177	ADA Attur, AEC Malliakarai	Salem
163	828	ADA Attur, AEC Peddanaickanpalayam	Salem
164	793	ADA Attur, AEC Thalaivasal	Salem
165	1182	ADA Attur, AEC Veeraganur	Salem
166	934	ADA Attur, AEC Gangavalli	Salem
167	1217	ADA Attur, AEC Thammammpatty	Salem
168	1419	ADA Namakkal	Namakkal
169	1321	ADA Namakkal, AEC Namakkal	Namakkal
170	830	ADA Namakkal, AEC Senthamangalam	Namakkal
171	1064	ADA Namakkal, AEC Belukkurichi	Namakkal
172	885	ADA Namakkal, AEC Erumappatty	Namakkal
173	1065	ADA Namakkal, AEC Muthukappatty	Namakkal
174	937	ADA Namakkal, AEC Puthansandai	Namakkal
175	1063	ADA Namakkal, AEC Thirumalaippatty	Namakkal
176	827	ADA Paramathy, AEC Paramathy	Namakkal
177	825	ADA Paramathy, AEC Kabilar malai	Namakkal
178	1062	ADA Paramathy, AEC Solasiramani	Namakkal

S.N.	S.P. Unit N.	Address	District
179	826	ADA Paramathy, AEC Moganur	Namakkal
180	50	ADA Rasipuram, Rasipuram	Namakkal
181	935	ADA Rasipuram, AEC Rasipuram	Namakkal
182	831	ADA Rasipuram, AEC Namagirippettai	Namakkal
183	1272	ADA Rasipuram, AEC Mullukkurichi	Namakkal
184	822	ADA Rasipuram, AEC Vennandur	Namakkal
185	1123	ADA Rasipuram, AEC Atthanur	Namakkal
186	960	ADA Thiruchengodu, AEC Thiruchengodu	Namakkal
187	959	ADA Thiruchengodu, AEC Erayamangalam	Namakkal
188	972	ADA Thiruchengodu, AEC Pallipalayam	Namakkal
189	974	ADA Thiruchengodu, AEC Veppadai	Namakkal
190	975	ADA Thiruchengodu, AEC Mallasamuthiram	Namakkal
191	973	ADA Thiruchengodu, AEC Vaiyappamalai	Namakkal
192	958	ADA Thiruchengodu, AEC Elachipalayam	Namakkal
193	957	ADA Thiruchengodu, AEC Velagoundapatty	Namakkal
194	889	ADA Dharampuri, AEC Dharampuri	Dharampuri
195	880	ADA Dharampuri, AEC Nallampalli	Dharampuri
196	799	ADA Pennagaram, SSF Paparappatty	Dharampuri
197	1435	ADA Pennagaram, AEC Pennagaram	Dharampuri
198	1195	ADA Pennagaram, AEC Paparappatty	Dharampuri
199	888	ADA Palacode, AEC Palacode	Dharampuri
200	994	ADA Palacode, AEC Karimangalam	Dharampuri
201	563	ADA Harur, Harur	Dharampuri
202	1471	ADA Harur, AEC Harur	Dharampuri
203	1119	ADA Harur, AEC Morapur	Dharampuri
204	1162	ADA Harur, AEC Papireddipatty	Dharampuri
205	403	ADA Krishnagiri, Krishnagiri	Krishnagiri
206	1512	ADA Krishnagiri, SSF Agasippalli	Krishnagiri
207	1141	ADA Krishnagiri, AEC Krishnagiri	Krishnagiri
208	1131	ADA Krishnagiri, AEC Veppanappalli	Krishnagiri
209	1170	ADA Krishnagiri, AEC Bargur	Krishnagiri
210	1504	ADA Krishnagiri, Central Godown	Krishnagiri
211	1172	ADA Krishnagiri, AEC Pochampalli	Krishnagiri
212	1320	ADA Krishnagiri, AEC Kaverippattinam	Krishnagiri
213	1171	ADA Krishnagiri, AEC Arasampatty	Krishnagiri
214	1132	ADA Uttangarai, AEC Mathur	Krishnagiri

S.N.	S.P. Unit N.	Address	District
215	1012	ADA Uttangarai, AEC Uttangarai	Krishnagiri
216	595	ADA Hosur, Hosur	Krishnagiri
217	1445	ADA Hosur, AEC Hosur	Krishnagiri
218	1134	ADA Uttangarai, AEC Soolagiri	Krishnagiri
219	1144	ADA Denganikotta, AEC Kelamangalam	Krishnagiri
220	1224	ADA Denganikotta, AEC Rayakkotta	Krishnagiri
221	1057	ADA Denganikotta, AEC Thali	Krishnagiri
222	1505	Uthangarai Agricultural Producers Co-op Marketing Society, Uthangarai	Krishnagiri
223	411	Kumaran Hybrid Seeds, 467-B Kamarajanar Salai, Attrur	Salem
224	517	Sri Balamurugan Hybrid Seeds, 175, Narasingapuram, Attur	Salem
225	540	Rasi Seeds (P) Ltd., 273, Kamarajanar Salai, Attur	Salem
226	550	Ganesh Ginning Factory, 30/2, Paithur Road, Attur	Salem
227	573	Selvam Hybrid Seeds, 26-D Kamarajanar Road, Attur	Salem
228	581	Arun Seeds, Chinna punalvaasal Punalvaasal (Po) Gangavalli	Salem
229	601	Siva Sakthi Ginning Factory, Trichy Mani Rd Thedaavur (PO) Gangavalli	Salem
230	610	Essar Seeds, 188/D-2 Kamarajanar Road, Attur	Salem
231	1025	Ganesh Ginning Factory, 60/1-B, Deviakkurichi, Thalaivasal	Salem
232	1279	South India Seeds, Cuddalore Main Road, Kallanatham (PO) Attur	Salem
233	1434	J.S.Seeds, 1/A, Cuddalore Main Road, Attur	Salem
234	1408	Aruna Hybrid Seed, 23-24, Ramasamy Nagar Pethanaickanpalayam Post	Salem
235	1338	Balaji Seeds, 200/ 1-E East Main Road, Thukkanampatty, Mettur dam	Salem
236	1516	Balamurugan Ginning Factory, Idappadi Main Road, Konganapuram.	Salem
237	1474	Sri Subalakshmi Hybrid Seeds, Manthiri Kaadu, Masakkalippatt, Rasipuram	Namakkal
238	1194	SPIC Biotech Ltd, Kelamangalam Road, Hosur Cattle Farm(Po) Hosur	Krishnagiri
239	665	AEC Andanallur	Trichy
240	666	AEC Allithurai	Trichy
241	844	AEC Melakalkandarkottai	Trichy

S.N.	S.P. Unit N.	Address	District
242	752	AEC Lalgudi @ Poovalur	Trichy
243	652	AEC Manachanallur	Trichy
244	797	AEC Pullambadi	Trichy
245	869	AEC Musiri	Trichy
246	749	AEC Thottiam	Trichy
247	753	AEC Thathaiyangarpettai	Trichy
248	776	AEC Thuraiyur	Trichy
249	653	AEC Uppiliyapuram	Trichy
250	654	AEC Manapparai	Trichy
251	765	AEC Marungapuri	Trichy
252	744	AEC Vaiyampatti	Trichy
253	365	SPU Manachanallur,	Trichy
254	567	SSF, Neikhuppaipudur, Lalgudi,	Trichy
255	1751	Dean, Agrl.Engg College, Kumulur, Lalgudi	Trichy
256	1515	Dhanya Seeds, (P) Ltd, Sri Prakesh Rice Mill Complex, Ariyamangalam, Kattur	Trichy
257	1484	Shanmuga Farm Seeds, 3/139 Devi Rice Mill, Kattur	Trichy
258	1238	Multiplex (P) Ltd, 11/5, A.K.A. Rice Mill Complex, Ariyamangalam, Kattur	Trichy
259	518	ADA (Pulses), Vamban	Pudukkottai
260	1511	ADA (Pulses), Vamban	Pudukkottai
261	372	ADA (Pulses), Vamban	Pudukkottai
262	37	ADA, Annapannai, Kudumianmalai	Pudukkottai
263	37(1)	ADA, Annapannai, Kudumianmalai	Pudukkottai
264	1278	Professor & Head, N.P.R.C (1278) Vamban	Pudukkottai
265	672	AEC Pudukkottai	Pudukkottai
266	805	AEC Thiruvarangulam	Pudukkottai
267	806	AEC Vettanviduthi	Pudukkottai
268	670	AEC K. Puduppatti	Pudukkottai
269	750	AEC Subramaniyapuram	Pudukkottai
270	671	AEC Annavasal	Pudukkottai
271	673	AEC Thirumayam	Pudukkottai
272	751	AEC Arantangi	Pudukkottai
273	778	AEC Keeranur	Pudukkottai
274	803	AEC Aththani	Pudukkottai
275	807	AEC Karambakudi	Pudukkottai
276	811	AEC Alangudi	Pudukkottai

S.N.	S.P. Unit N.	Address	District
277	1029	AEC Kunandarkovil	Pudukkottai
278	1477	AEC Ponamaravthi	Pudukkottai
279	804	AEC Kandarvakkottai	Pudukkottai
280	651	AEC Viralimalai	Pudukkottai
281	669	AEC Arimalam	Pudukkottai
282	872	ADA (SOSF), Vellalaviduthi	Pudukkottai
283	1231	Seed Processing Unit, Thirumanur,	Perambalur
284	845	AEC Perambalur	Perambalur
285	1466	AEC Ammapalayam	Perambalur
286	1023	AEC Veppur	Perambalur
287	777	AEC Veppanthattai	Perambalur
288	846	AEC Alathurgate	Perambalur
289	911	AEC Ariyalur	Perambalur
290	1084	AEC Thirumanur	Perambalur
291	1024	AEC Sendurai	Perambalur
292	767	AEC Jayankondam	Perambalur
293	775	AEC T. Palur	Perambalur
294	1444	AEC Andimadam	Perambalur
295	193	Thambu Udayar Sons, Perambalur	Perambalur
296	603	Madura Ginning Factory, Perambalur	Perambalur
297	122	SSF Inungur, Karur	Karur
298	650	AEC Velayuthampalayam	Karur
299	809	AEC Thanthoni	Karur
300	675	AEC K. Paramathi	Karur
301	674	AEC Aravakurichi	Karur
302	938	AEC Kulithalai	Karur
303	941	AEC Krishnaryapuram	Karur
304	1034	AEC Kadavur @ Tharagampatti	Karur
305	801	AEC Thogaimalai	Karur
306	79	SSF Sakkottai, Kumbakonam	Thanjavur
307	360	Kattuthottam SPU, Thanjavur	Thanjavur
308	578	SSF Sakkottai,	Thanjavur
309	442	Sakkottai Ginning Unit, SSF Sakkottai	Thanjavur
310	694	AEC Thanjavur	Thanjavur
311	1098	AEC Vallam	Thanjavur
312	1210	AEC Thirukattupalli	Thanjavur

S.N.	S.P. Unit N.	Address	District
313	1497	AEC Budalur	Thanjavur
314	684	AEC Thiruvaiyaru	Thanjavur
315	783	AEC Thiruchittambalam	Thanjavur
316	1189	AEC Ganapathyagragaram	Thanjavur
317	903	AEC Thirukarukavoor	Thanjavur
318	862	AEC Pattukottai	Thanjavur
319	759	AEC Pookollai	Thanjavur
320	780	AEC Peravurani	Thanjavur
321	925	AEC Orathanadu	Thanjavur
322	784	AEC Madukkur	Thanjavur
323	782	AEC Thiruvidaimaruthur	Thanjavur
324	788	AEC Kanjanoor	Thanjavur
325	1427	Chandrakala SPU, Thanjavur	Thanjavur
326	1502	Sri Subramania Ginning Mills, Kumbakonam	Thanjavur
327	162	Sri Swaninatha Ginning Mills, Kumbakonam	Thanjavur
328	1306	V.V. SPU, Ailvalam, Pattukottai	Thanjavur
329	1307	Arul SPU, Gandhiji Road, Thanjavur	Thanjavur
330	1259	Anandha Processing Unit, 61/1 vps mill, Vadasery Road, Sorappallam, Thanjavur	Thanjavur
331	1097	Kasi Muthu Kumara Swamy, Seed Processing Unit, Thirupanandhal	Thanjavur
332	41	Thiruvarur SPU, Pallivasal St, Thiruvarur	Thiruvarur
333	381	SSF SPU, Thiyagarajapuram, Thiruvarur	Thiruvarur
334	498	SSF SPU, Kanchikudikadu, Thiruvarur	Thiruvarur
335	497	SSF SPU, Nedumbalam, Thiruvarur	Thiruvarur
336	246	SSF SPU, Orathur, Thiruvarur	Thiruvarur
337	773	AEC Thiruvarur	Thiruvarur
338	759	AEC Alangudi	Thiruvarur
339	859	AEC Needamangalam	Thiruvarur
340	877	AEC Vaduvoor	Thiruvarur
341	863	AEC Paravakkottai	Thiruvarur
342	1270	AEC Vadapathimangalam	Thiruvarur
343	1209	AEC Valloor	Thiruvarur
344	902	AEC Kudavasal	Thiruvarur
345	1014	AEC Nanilam	Thiruvarur
346	899	AEC Edaiyur	Thiruvarur
347	900	AEC Thiruthuraipoondi	Thiruvarur

S.N.	S.P. Unit N.	Address	District
348	345	TANFED SPU, Meenakshi amman Kovil St, Mannargudi	Thiruvarur
349	1159	Consolidated SPU, Karuppur, Thiruvarur	Thiruvarur
350	631	Nathan SPU, South St, Paravakkottai,	Thiruvarur
351	80	SSF SPU, Nagamangalam	Nagapattinam
352	34	SSF SPU, Tharanganpadi, Nagapattinam	Nagapattinam
353	918	AEC Keevalur,	Nagapattinam
354	1099	AEC Sirgali	Nagapattinam
355	850	AEC Kollidam	Nagapattinam
356	963	AEC Thirumarugal	Nagapattinam
357	904	AEC Nagapattinam	Nagapattinam
358	1423	AEC Velanganni	Nagapattinam
359	1306	K.S.O.SPU, Thiruvarur Main Rd, Manganallur	Nagapattinam
360	557	SSF, Thenkasi Rd, Devadanam.	Virudhunagar
361	404	Govt SPU, A.Ramalingapuram, Sattur	Virudhunagar
362	358	Govt SPU, Gopalapuram, Aruppukottai	Virudhunagar
363	368	Lakshmi Farm Seeds, 16 Main Rd, Mettupatty, Sattur	Virudhunagar
364	373	Andal Fertilisers, 235-A Krishnankoil St, Srivilliputhur	Virudhunagar
365	1514	K.B. Bharath Singh, 146/D1,D2, Andalpuram, Rajapalayam	Virudhunagar
366	647	AEC, Pullalakottai Road, Virudhunagar	Virudhunagar
367	679	AEC, Panchayat Union Complex, Sattur	Virudhunagar
368	678	AEC, Panchayat Union Complex, Sivakasi	Virudhunagar
369	723	AEC, Panchayat Union Complex, Vembakottai	Virudhunagar
370	1296	AEC, Panchayat Union Complex, Srivilliputhur	Virudhunagar
371	698	AEC, Panchayat Union Complex, Rajapalayam	Virudhunagar
372	710	AEC, Panchayat Union Complex, Watrap	Virudhunagar
373	882	AEC, Panchayat Union Complex, Kariapatty	Virudhunagar
374	696	AEC, Panchayat Union Complex, Aruppukottai	Virudhunagar
375	730	AEC, Narikudi Road, Tiruchuli	Virudhunagar
376	731	AEC, Panchayat Union Complex, M.Reddiapatty	Virudhunagar
377	747	AEC, Panchayat Union Complex, Narikudi	Virudhunagar
378	1469	N.R. Krishnamoorthyraja, 200/163, Cotton Market, Rajapalayam	Virudhunagar
379	1406	Balaiah, Mamsapuram, Srivilliputhur	Virudhunagar
380	1404	G.Manoharan, Gopalapuram, TNC, Alangulam,	Virudhunagar

S.N.	S.P. Unit N.	Address	District
		Sivakasi	
381	1212	Jayasree Seeds, 1/45, West Street, Sippiparai, Sivakasi	Virudhunagar
382	1288	Selvarani Seeds, 22, Middle Street, Uppathur, Sattur	Virudhunagar
383	427	Govt SPU, Thirukosamangai, Ramnad	Ramanathapuram
384	561	Govt SPU, Panchayat Union Complex, Paramakudi	Ramanathapuram
385	1424	AEC, Panchayat Union Complex, Ramnad	Ramanathapuram
386	838	AEC, Panchayat Union Complex, Tiruppulani	Ramanathapuram
387	833	AEC, Panchayat Union Complex, Madurai Road, Chatrakudi	Ramanathapuram
388	711	AEC, Panchayat Union Complex, Thondi Road, Tiruvadanai	Ramanathapuram
389	714	AEC, Panchayat Union Complex, Kottaimedu, Kamuthi	Ramanathapuram
390	832	AEC, Panchayat Union Complex, Ramnad Road, Nainarkoil	Ramanathapuram
391	1015	AEC, Panchayat Union Complex, Mudukulathur	Ramanathapuram
392	1277	AEC, Panchayat Union Complex, Kadaladi	Ramanathapuram
393	1286	AEC, Tuticorin Road, Sayalkudi	Ramanathapuram
394	1330	AEC, Panchayat Union Complex, Paramakudi	Ramanathapuram
395	704	AEC, Mandabam Road, Uchipuli	Ramanathapuram
396	364	ADA, Thondi Road, Sivaganga	Sivaganga
397	1300	Govt SPU, Panchayat Union Complex, Devakottai	Sivaganga
398	914	AEC, Opp Panchayat Union, Madurai Road, Thiruppuvanam,	Sivaganga
399	1030	AEC, Panchayat Union Complex, Singampunari	Sivaganga
400	915	AEC, Panchayat Union Complex, Thiruppathur	Sivaganga
401	996	AEC, Panchayat Union Complex, Sivaganga	Sivaganga
402	1048	AEC, Panchayat Union Complex, Kalayarkoil	Sivaganga
403	837	AEC, Panchayat Union Complex, Devakottai	Sivaganga
404	842	AEC, Panchayat Union Complex, Kannangudi	Sivaganga
405	847	AEC, Panchayat Union Complex, Sakkottai	Sivaganga
406	1333	AEC, Panchayat Union Complex, Kallal	Sivaganga
407	1328	AEC, Panchayat Union Complex, Manamadurai	Sivaganga
408	987	AEC, Panchayat Union Complex, ilayangudi	Sivaganga
409	1332	Asst Director of Horticulture, Hospital Road, Devakottai	Sivaganga
410	447	AO (F.A) Karairuppu	Tirunelveli
411	555	AEC Ambai	Tirunelveli

S.N.	S.P. Unit N.	Address	District
412	473	AEC Thenkasi	Tirunelveli
413	1503	Sankarankovil Agri Producers Co-operative Mkt Society Ltd.Sankarankovil	Tirunelveli
414	645	S. Mahesh Seeds, Puliyankudi	Tirunelveli
415	362	Selva Seeds, Puliyankudi	Tirunelveli
416	474	Santhanakrishna Ginning Factory, Sennikulam	Tirunelveli
417	1416	S. Subbaiya, Ayakudi	Tirunelveli
418	1362	Govindhan, Subbaiyapuram	Tirunelveli
419	1080	V. Anthonysamy, Puliyankudi	Tirunelveli
420	1482	Santhosh Farm Seeds, Vasudevanallur	Tirunelveli
421	1486	Samy Seed Farm, Puliyankudi	Tirunelveli
422	748	ADO, Tirunelveli	Tirunelveli
423	1275	ADO, Manur	Tirunelveli
424	1191	ADO, Gangaikondon	Tirunelveli
425	821	ADO, Palayankottai	Tirunelveli
426	1252	ADO, Muneerpallam	Tirunelveli
427	1464	ADO, Tirunelveli Junction	Tirunelveli
428	1495	Central Seed Godown, Palayankottai	Tirunelveli
429	1173	ADO, Nanguneri	Tirunelveli
430	668	ADO, Moolaikaraipatti	Tirunelveli
431	840	ADO, Kallakad	Tirunelveli
432	1180	ADO, Yervadi	Tirunelveli
433	1184	ADO, Padmaneri	Tirunelveli
434	853	ADO, Valliyur	Tirunelveli
435	817	ADO, Palavur	Tirunelveli
436	1096	Ado, Radhapuram	Tirunelveli
437	1250	ADO, Thisaiyanvilai	Tirunelveli
438	1126	ADO, Ambasamudram	Tirunelveli
439	722	ADO, Cheranmadevi	Tirunelveli
440	1215	ADO, Veeravanallur	Tirunelveli
441	907	ADO, Mukkudal	Tirunelveli
442	839	ADO, Kadaiyam	Tirunelveli
443	473	ADO, Thenkasi	Tirunelveli
444	890	ADO, Kadaiyanallur	Tirunelveli
445	655	ADO, Pavurchatram	Tirunelveli
446	874	ADO, Alankulam	Tirunelveli

S.N.	S.P. Unit N.	Address	District
447	771	ADO, Chenkkottai	Tirunelveli
448	770	ADO, Ayakudi	Tirunelveli
449	700	ADO, Vasudevanallur	Tirunelveli
450	699	ADO, Puliankudi	Tirunelveli
451	701	ADO, Sivakiri	Tirunelveli
452	727	ADO, Mullikulam	Tirunelveli
453	1185	ADO, Sankarankoil	Tirunelveli
454	721	ADO, Kuruvikulam	Tirunelveli
455	720	ADO, Thiruvenkadam	Tirunelveli
456	867	ADO, Melanilithanallur	Tirunelveli
457	779	ADO, Sernthamaram	Tirunelveli
458	108	SPU, Killikulam	Tuticorin
459	431	ADO, Kovilpatti	Tuticorin
460	1061	Kovilpatti Co-operative Agri Marketing Society Yettaiyapuram	Tuticorin
461	1246	Scientifically Select Seeds, Kovilpatti	Tuticorin
462	1258	Chandiya Farm Service, Kovilpatti	Tuticorin
463	1463	Sri Srinivasa Ginning Factory, Kilayeeral	Tuticorin
464	1310	Thilagam ginning Factory, Kilayeeral	Tuticorin
465	1442	Langashyaar ginning Factory, Kilayeeral	Tuticorin
466	1080	Jayalakshmi Ginning Factory, Karisalkulam	Tuticorin
467	1508	T.L. Lakshmana Perumal Ginning Factory Kovilpatti	Tuticorin
468	658	ADO, Pudukkottai	Tuticorin
469	660	ADO, Ottapidaram	Tuticorin
470	661	ADO, Ottanatham	Tuticorin
471	854	ADO, Yettaiyapuram	Tuticorin
472	657	ADO, Kayatharu	Tuticorin
473	841	ADO, Kazhukumalai	Tuticorin
474	646	ADO, Thiruvaikundam	Tuticorin
475	928	ADO, Seithunganallur	Tuticorin
476	663	ADO, Vallanadu	Tuticorin
477	729	ADO, Sonakanvilai	Tuticorin
478	724	ADO, Thenthiruperai	Tuticorin
479	905	ADO, Nazareth	Tuticorin
480	725	ADO, Yudankudi	Tuticorin
481	726	ADO, Sathankulam	Tuticorin

S.N.	S.P. Unit N.	Address	District
482	659	ADO, Villathikulam	Tuticorin
483	667	ADO, Pudur	Tuticorin
484	1260	Selvakani SPU, Kovilpatti	Tuticorin
485	375	Madurai Chemical Company, Kovilpatti	Tuticorin
486	1251	Sri Vishnu Seeds, Yettaiuapuram	Tuticorin
487	1253	Keerthi Seeds, Iyan Bommaiyapuram	Tuticorin
488	1496	Suganya Traders,	Tuticorin
489	40	Agricultural Officer (F.A) Thiruppathisaram	Kanniakumari
490	930	ADO, Agastheeshwaram	Kanniakumari
491	897	ADO, Rajakkamangalam	Kanniakumari
492	892	ADO, Boothapandi	Kanniakumari
493	988	ADO, Melpuram	Kanniakumari
494	967	ADO, Killiyur	Kanniakumari
495	895	ADO, Munsirai	Kanniakumari
496	718	ADO, Koliporvilai	Kanniakumari
497	715	ADO, Kurunthankodu	Kanniakumari
498	719	ADO, Thiruvattar	Kanniakumari
499	92	ADA, Vadipatti, SPU, Chozhavanthan	Madurai
500	606	T. Stanes & Co Pvt Ltd., 139/2 Aruppukottai main road, Avanipuram, Madurai.	Madurai
501	1329	K. Shanmuga Seeds, Thottiyapatti, Madurai	Madurai
502	93	Agricultural Officer (F.A) Govt. Seed Farm Melur, Madurai	Madurai
503	355	National Seeds Corporation, Kappalur	Madurai
504	1254	Southern Hi-tech Seed and Horticulture Farm Varichiyur	Madurai
505	1311	Jayaram Seeds Farm, 67. Vidathakulam Road Thirumangalam	Madurai
506	1324	Agricultural College and Research Centre, Madurai	Madurai
507	1492	Production Officer, AEW, 120. Siddco Campus Kappalur, Madurai	Madurai
508	961	ADO, Thiruvparangundram	Madurai
509	781	ADO, Madurai East	Madurai
510	1418	ADO, T. Vadipatti	Madurai
511	815	ADO, Kallikudi	Madurai
512	814	ADO, T. Kallupatti	Madurai
513	677	ADO, Thirumangalam	Madurai

S.N.	S.P. Unit N.	Address	District
514	818	ADO, Palamedu	Madurai
515	819	ADO, Anaganallur	Madurai
516	709	ADO, Melur	Madurai
517	800	ADO, Usilampatti	Madurai
518	1401	ADO, Sedapatti	Madurai
519	962	ADO, Chellampatti	Madurai
520	848	ADO, Kottampatti	Madurai
521	1331	Thirupathi ginning Factory, Melakkottai Thirumangalam	Madurai
522	1501	NSK.Ginning and Oil Factory, Vidathakulam Road,Thirumangalam	Madurai
523	1176	C. Periakaruppan, 302, Karunkalakudi, Melur	Madurai
524	1426	Agricultural Officer (F.M) Govt. Seed Farm Keelgudalur, Uthamapalayam	Theni
525	626	Professor & Head, A.R.C .Vaigaidam, Periakulam	Theni
526	1257	Balaji Farm Seeds, 223, Rathinam Nagarm, Theni	Theni
527	1461	Ganesh Seeds, 7. Periakulam Road, Bommaiyagoundenpatti	Theni
528	1226	Gobikrishna Seeds, Siddco No.4, Industrial Estate	Theni
529	910	SRR.Agri Business Associates, Rasingapuram, Bodi	Theni
530	1449	SRR Agri Business Associates, 20/15-5-14.F, Bothi-puram Road, Theni	Theni
531	792	ADO, Theni	Theni
532	812	ADO, Silamalai	Theni
533	754	ADO, Vadugapatti	Theni
534	1289	ADO, Devadhanapatti	Theni
535	702	ADO, Aandipatti	Theni
536	851	ADO, Kadamalaigundu	Theni
537	1325	ADO, Uthamapalayam	Theni
538	1323	ADO, Kambam	Theni
539	1400	ADO, Chinnamanur	Theni
540	910	ADO, Bodinaickanur	Theni
541	548	A-1254 Theni Co-operative Marketing Society Allinagarm	Theni
542	320	P. Masilamani, D- No.30 T. Vadipatti Erumalainaickenpatti Post	Theni
543	440	Sriram Seeds, 25, Nethaji Road, NRT. Nagar	Theni
544	664	T.R. Dinakaran, SRM.Press Factory Periakulam Road,	Theni

S.N.	S.P. Unit N.	Address	District
		Theni	
545	325	ADA, Palani	Dindigul
546	756	ADO, Ottanchathiram	Dindigul
547	705	ADO, Reddiyarchathiram	Dindigul
548	755	ADO, Dindigul	Dindigul
549	813	ADO, Vedachendur	Dindigul
550	916	ADO, Natham	Dindigul
551	713	ADO, Athur (Eruppu) Sempatti	Dindigul
552	929	ADO, Vadamadurai	Dindigul
553	912	ADO, Nilakkottai	Dindigul
554	772	ADO, Vathalagundu	Dindigul
555	820	ADO, Sanarpatti	Dindigul
556	703	ADO, Palani	Dindigul
557	1271	ADO, Gujiliamparai	Dindigul
558	708	ADO, Thoppampatti	Dindigul
559	1425	National Hort.Res. & Pro. Foundation 14A, Krishnarao St. Pandian Nagar,	Dindigul
560	1059	Pandiyan Seeds, 6-6/16, Thinakaran St. Pattiveeranpatti	Dindigul
561	536	K.Gopalsamy, T.K.N. Pudur, Aayakudi Post	Dindigul
562	1249	M/s. Balamurugan Ginning Factory, Chathrapatti	Dindigul
563	1422	M.S.N.P.T. Jaisankar, Pandian Seed Farm Pattiveeranpatti	Dindigul
564	89	SPU. Vasur, Polur TK	T.V.Malai
565	1315	SPU. Athiyanthal,T.V. Malai	T.V.Malai
566	1479	SPU, Cheyar	T.V.Malai
567	1153	ADO, Vandhavasi	T.V.Malai
568	1157	ADO, Keelkodungalaur	T.V.Malai
569	1465	ADO, Peranamallur	T.V.Malai
570	1446	ADO, Thurinjapuram	T.V.Malai
571	1193	ADO, Thellar	T.V.Malai
572	1194	ADO, Thesur	T.V.Malai
573	1113	ADO, Polur	T.V.Malai
574	1112	ADO, Chanthavasal	T.V.Malai
575	1440	ADO, Chethuppattu	T.V.Malai
576	1111	ADO, Kalambur	T.V.Malai
577	1136	ADO, Kalasapakkam	T.V.Malai

S.N.	S.P. Unit N.	Address	District
578	1137	ADO, Aathamangalam	T.V.Malai
579	1115	ADO, T.V. Malai	T.V.Malai
580	1286	ADO, Vembakkam	T.V.Malai
581	1106	ADO, Keelpenathur	T.V.Malai
582	1488	ADO, Kommanthal	T.V.Malai
583	1223	ADO, Dhachampattu	T.V.Malai
584	1454	ADO, Devikapuram	T.V.Malai
585	1274	ADO, Periakulam	T.V.Malai
586	1455	ADO, Kannamangalam	T.V.Malai
587	676	ADO, Padhupalayam	T.V.Malai
588	1459	ADO, Chengam	T.V.Malai
589	1456	ADO, Aarani (West)	T.V.Malai
590	951	ADO, Thandurampattu	T.V.Malai
591	1458	ADO, Payichal	T.V.Malai
592	1470	ADO, Aarani	T.V.Malai
593	1336	ADO, Vettalavalam	T.V.Malai
594	1428	ADO, Anaakkavur	T.V.Malai
595	1453	SSF, Valavachanur	T.V.Malai
596	1139	SSF, Athiyanthal	T.V.Malai
597	618	Area Manager, Cheyar (Iruppu) Thuli	T.V.Malai
598	1242	Area Manager, Cheyar (Iruppu) Thumbai	T.V.Malai
599	1481	Area Manager, T.V.Malai (Iruppu) Venkikal	T.V.Malai
600	1420	Karthik Seeds, 103. Muthu St.Thandarampattu	T.V.Malai
601	49	SPU. Kalputhur, Katpadi	Vellore
602	363	SPU, Avarakarai,Valaja	Vellore
603	1230	Hort. Officer, Vellore	Vellore
604	1078	ADO, Vellore	Vellore
605	1160	ADO, Oosoor	Vellore
606	1006	ADO, Katpadi	Vellore
607	1000	ADO, Gudiyatham	Vellore
608	1207	ADO, Chakkaramallur	Vellore
609	1208	ADO, Arcot	Vellore
610	1441	ADO, Madhanur	Vellore
611	1433	ADO, Aalankalam	Vellore
612	1432	ADO, Peranampattu	Vellore
613	1155	ADO, Anaikattu	Vellore

S.N.	S.P. Unit N.	Address	District
614	1206	ADO, Thimiri	Vellore
615	1269	ADO, Kalavai	Vellore
616	1077	ADO, Kaniyampadi	Vellore
617	1151	ADO, K.V.Kuppam	Vellore
618	1237	ADO, Cholingar	Vellore
619	953	ADO, Valaja	Vellore
620	1001	ADO,. Thiruppathur	Vellore
621	1439	ADO, Natrampalli	Vellore
622	1284	ADO, Kanthili (Iruppu) Thiruppathur	Vellore
623	1429	ADO, Nemili	Vellore
624	1430	ADO, Panappakkam	Vellore
625	1071	ADO, Jolarpettai (Iruppu) Kodiyur	Vellore
626	1476	ADO, Kaverippakkam	Vellore
627	1431	ADO, Arakkonam	Vellore
628	1421	ADO, Melpadi	Vellore
629	1074	State Oilseed Farm, Novlock	Vellore
630	255	SPU, Panchupet, Kancheepuram	Kancheepuram
631	48	SPU, Kattankulathur	Kancheepuram
632	1282	SPU, Thirukalukundram	Kancheepuram
633	17	SPU, Madhurandhagam	Kancheepuram
634	1052	SPU, Achsiruppakkam	Kancheepuram
635	402	SPU, Kundrathur	Kancheepuram
636	1229	ADO, Sirukaveripakkam	Kancheepuram
637	1221	ADO, Pounjur	Kancheepuram
638	1234	ADO, Valajabath	Kancheepuram
639	1007	ADO, Thirupporur	Kancheepuram
640	1100	ADO, Uthiramerur	Kancheepuram
641	1148	ADO, Sriperumpudur	Kancheepuram
642	1129	ADO, Padappai	Kancheepuram
643	1102	ADO, Chittlapakkam	Kancheepuram
644	1054	ADO, Perumbakkam	Kancheepuram
645	896	ADO, Chengalpet	Kancheepuram
646	1220	ADO, Chithamur	Kancheepuram
647	1025	SOSF, Musaravakkam	Kancheepuram
648	1281	SPU, Kolundalur	Thiruvallur
649	359	SPU, Cholavaram	Thiruvallur

S.N.	S.P. Unit N.	Address	District
650	391	SPU, Meenjur	Thiruvallur
651	1491	SPU, Aagurkiramam, Thiruthani	Thiruvallur
652	978	ADO, Kadambathur	Thiruvallur
653	1043	ADO, Ellapuram	Thiruvallur
654	1069	ADO, Puzhalal	Thiruvallur
655	1017	ADO, Eigadu	Thiruvallur
656	1198	ADO, Vellavedu	Thiruvallur
657	1197	ADO, Poonthamalli	Thiruvallur
658	1265	ADO, Poondi	Thiruvallur
659	1127	ADO, Kanakkammachathiram	Thiruvallur
660	1228	ADO, Thiruthani	Thiruvallur
661	1010	ADO, Pondhavakkam	Thiruvallur
662	1314	ADO, Pethatturpet	Thiruvallur
663	1490	ADO, Aarani	Thiruvallur
664	1219	ADO, Ambathur	Thiruvallur
665	977	ADO, Pallipattu	Thiruvallur
666	1047	ADO, C.N.Poondi	Thiruvallur
667	1044	ADO, R.K.Pettai	Thiruvallur
668	1450	ADO, Gumidipoondi	Thiruvallur
669	1402	ADO, Pattra Perumpudhur	Thiruvallur
670	1438	ADO, Poochiyathupedu	Thiruvallur
671	1489	ADO, Uthukkottoai	Thiruvallur
672	1101	ADO, Thennaithaimara thottam, Putlur	Thiruvallur
673	1499	NSC, 22C. Siddco Industrial Estate, Ambathur	Thiruvallur
674	1240	Dilip Seeds, Eliampedukiramam, Ponneri	Thiruvallur
675	1045	Srisidhivinayakar Farm, Pandikavalur, Ponneri	Thiruvallur
676	222	Srisidhivinayakar Farm, Pandikavalur, Ponneri	Thiruvallur
677	83	SPU, Kallakurichi	Villupuram
678	260	SPU, Eruvelpattu	Villupuram
679	47	SPU, Ka Kuppam	Villupuram
680	1032	ADO, Kanai	Villupuram
681	982	ADO, Mugaiyur(E) Manampoondi	Villupuram
682	1005	ADO, Kallakurichi	Villupuram
683	1493	ADO, Sankarapuram	Villupuram
684	924	ADO, Olangur (E) Saram	Villupuram
685	946	ADO, Murunkeri	Villupuram

S.N.	S.P. Unit N.	Address	District
686	898	ADO, Chenji	Villupuram
687	1462	ADO, Vallam	Villupuram
688	1146	ADO, Anandhapuram	Villupuram
689	1147	ADO, Avalurpettai	Villupuram
690	873	ADO, Melmalaianur	Villupuram
691	871	ADO, Kooteripattu	Villupuram
692	948	ADO, Thirukovilur	Villupuram
693	1110	ADO, Tindivanam	Villupuram
694	940	ADO, Ariyalur	Villupuram
695	786	ADO, Thiruvennainallur	Villupuram
696	1040	ADO, Vikkravandi	Villupuram
697	1028	ADO, Ulunthurpet	Villupuram
698	970	ADO, Kandamangalam	Villupuram
699	943	ADO, Thirunavalur	Villupuram
700	1004	ADO, Chinnasalem	Villupuram
701	969	ADO, Vanur	Villupuram
702	1507	ADO, Marakkanam	Villupuram
703	1248	Rasi Seeds, Thachur Village, Kallakurichi	Villupuram
704	1255	Prithivi Ginning Factory	Villupuram
705	1317	Lakshmi Seeds Company, Kilaiyur Village Thirukovilur	Villupuram
706	1305	Meenakshi Seeds, Sirumadurai, T.Kovilur	Villupuram
707	568	Area Manager, A.E.Wing, Neiveli	Cuddalore
708	628	Area Manager, A.E.Wing, Virudhachalam	Cuddalore
709	65	SPU, Vandurayanpattu, Chidhambaram	Cuddalore
710	308	SPU, Miralur, Chidhambaram	Cuddalore
711	860	ADO, Kattumannarkoil	Cuddalore
712	944	ADO, Ammapettai	Cuddalore
713	945	ADO, Kirapalayam	Cuddalore
714	876	ADO, Panruti	Cuddalore
715	917	ADO, Annakiramam,	Cuddalore
716	947	ADO, Cuddallore	Cuddalore
717	861	ADO, Virudhachalam	Cuddalore
718	785	ADO, Melpuvanakiri	Cuddalore
719	939	ADO, Kurinjipaddi	Cuddalore
720	1452	ADO, Thittakudi	Cuddalore

S.N.	S.P. Unit N.	Address	District
721	1407	ADO, Chidhambaram	Cuddalore
722	1003	ADO, Kammapuram	Cuddalore
723	1443	ADO, Nallur	Cuddalore
724	1494	Sri Bharathi Hybrid Seeds, Komangalam Virudhachalam	Cuddalore
725	1513	Garuda Seeds, Mettupalayam	Cuddalore
726	1318	Ganga Seeds, Aandalmullipallam	Cuddalore

Orissa State Seed Certification Agency

	Name of the Industry & Location	Distt
1	OSSC Bargarh	Bargarh
5	OUAT Farm Gambahripalli	Bargarh
7	Govt.Samakhunta SPP Samakunta	Mayurbahanj
8	Govt Sadar Farm,Keonjhargarh	Keonjhar
26	OSSC Randapalli	Koraput
27	OSSC Ltd SPP Behrampur,Langipally	Ganjam
28	OSSC Industrial Estate Balasore	Balasore
29	OUAT Instructional Farm Bhubaneshwer	Khurda
30	OSSC Ltd Chandaka	Khurda
32	Govt Seed Farm Chakuli	Chakuli
34	Govt Paljihar Farm Paljihar	Ganjam
36	OUAT NAPR Mahisapat	Dhenkanal
37	OUAT, Ranital Plant ,Ranital Bhadrak	Balasore
38	Govt Dhanel Farm Dhanei	Ganjam
41	OSSC Cold Storage Cuttack	Cuttack
43	ASPU Dunguripalli	Bolangir
47	OSSC Keonjhar SPP Keonjhar	Baripada
48	OSSC RCMS Kenjohar	Mayurbahanj
49	Govt Seed Farm lachida	bargarh
50	OSSC DAO Campus Narayanpur	Koraput
51	OSSC Sadar Farm SNG Sunder garh	Sundargarh
52	Govt SPP Bargaon	Sundargarh
53	Govt Seed Farm Banki	Cuttack
54	Govt. Baharda Seed Farm, Baharda	Balasore
55	Govt. Agarpada Seed Farm Seulagadia, Bhadrak	Balasore
56	OSSC, Bhadrak Seeds Farms, Haladidiha, Bhadrak	Balasore

	Name of the Industry & Location	Distt
57	OSSC, Gohira SPP Gohira	Keonjhar
58	OSSC Bhawanipatna, Arkabahal	Kalahandi
59	OSSC Ltd. Chandaka	Khurda
60	Bijaya Seed Industry Bhatra	Sambalpur
62	Govt PS Nayagarh	Nayagarh
64	Govt SPP Farm Tirtol	Tirtol
65	Govt Agri Farm Athgarh	Athgarh
67	Govt Olans Farm Olans	Puri
69	Govt R.E farm Bolangir	Bolangir
70	Govt Agri. Farm Birmaharajpur	Sonipur
71	Govt Agri farm Rampur	Bolangir
72	Govt. Baisinga Agriculture Farm, Basinga Distt	Mayurbahanj
73	OSSC, Dhanapana Agri. Farm Dhanapana	Mayurbahanj
74	Govt. Dalki Agri Farm, Dalki	Mayurbahanj
75	Govt. Sandeuli Agri Farm, Sandeuli	Mayurbahanj
77	Hindusthan Seeds Maingoung	Bolangir
79	OSSC Dabugong	Nabarangpur
80	OSSC SPP Umarcoat	Nabarangpur
82	OSSC SPP Agri Farm Kotpad	Koraput
83	Govt SPP Boriguma	Koraput
84	OSSC Agri. Farm Methli	Malkangiri
85	Govt SPP Narayanpatna	Koraput
86	Govt SPP Sangam Keonjhar	Baripada
88	Govt Seed Farm Damasala	Dhenkanal
89	Govt Seed Farm Gondia	Dhenkanal
90	OSSC, Dhamnager Plant, Pirpipur, Bharak	Balasore
92	OSSC Deogarh	Deogarh
93	OSSC Lambal Sambalpur	Sambalpur
96	Govt Bhanjanagar Farm Bhanjanagar	Ganjam
97	Govt Agri farm Khariar	Nauapara
98	Govt Agri farm Nuapara	Nuapara
100	Govt SPP Farm Kuliposh	Sundargarh
101	Govt Agri farm Pathangarh	Bolangir
102	Govt Seed Farm Pallahara	Anugul
103	Govt Seed Farm GK Pur Gatikrishnapur	Anugul
107	Kissan Seeds Pvt Ltd Rehanja	Balasore

	Name of the Industry & Location	Distt
108	Pragati Seeds Paralakhemundi	Gajapatti
109	OSSC Cold Storage Cuttack	Cuttack
111	OSSC Arkabahal Farm Arkabahal	Bolangir
112	OUAT RRTTS Chipilima	Sambalpur
113	Chandrashekar Agro Tech Staion Bazar Dhenkanal	Cuttack
114	Samaleshwari Seeds Gosala	Sambalpur
115	OSSC Veg. SPP Chandaka	Khurda
116	Anima Seeds Bohindamunda	Sambalpur
117	Amulya Seeds Charbahal	Kalahandi
119	Majhi Seeds Sikshakapra	Anugul
120	Sai Agro Products Talapatia	Jharasuguda
125	OSSC Barikel	Bargarh
127	Sood Seeds Khariar	Nuapara
128	Om Shiva Seeds Dubari	Jajpur
129	Chandrashekar Agro Tech (Ginning) Dhenkanal	Dhenkanal
130	Bibekhananda Seeds Attabira	Bargarh
131	Sood Seeds Khariar	Nuapara
132	Jagannath seeds Baladiabandha	Dhenkanal
133	Orissa Seeds Bhatra	Sambalpur
134	OSSC Randapalli	Koraput
135	Om Seeds Amapalli Bhidar Naupalli	Bargarh
136	Birtia seeds Bargarh	Bargarh
137	Bhulaxmi Seeds Padhanpalli	Bargarh
138	Live & Let Live Seed Processing Olant Nuagarh Redhua	J.S .Pur
139	National Seeds Corporation BBSR Bhubaneshwer	Khurda
140	Vivakananda Seeds Attabira	Bargarh
141	Bharat Seeds Anilapali, Dungripally	Sambalpur
142	OUAT CRS OUAT BBSR Bhubaneshwer	Khurda
143	Govt SPP Farm Bargaon	Sambalpur
144	Sansar Agropol Bhagabatipur	Puri

Chapter - 19

Notified Varieties and their Notification Status from 1997

S.No	Crop/Variety	S.O.	Date
	Bajra (Pearl Millet) (*Pennisetum americanism*)		
1	MLBH-285 (MH-518)	360/E	01/05/1997
2	NANDI-30 (MH-515)	360/E	01/05/1997
3	CZ-IC-923 (MP-258)	360/E	01/05/1997
4	PUSA BAJRI-266 (MP-266)	360/E	01/05/1997
5	SABURI (MH-483, RHRBH-8924)	360/E	01/05/1997
6	X-6 (MH-140)	360/E	01/05/1997
7	RBH-30	360/E	01/05/1997
8	GHB-183	360/E	01/05/1997
9	X-7	647/E	09/09/1997
10	JKBH-26 (MH-595)	662/E	17/09/1997
11	GHB-316 (MH-670)	662/E	17/09/1997
12	ANANTA (APS-1)	662/E	17/09/1997
13	GK-1004 (MH-662)	401/E	15/05/1998
14	7686 (MH-643*M-631)	401/E	15/05/1998
15	PAC-903 (ICI-903, MH-552)	401/E	15/05/1998
16	PROAGRO-1 (FMH-3) (FORAGE)	401/E	15/05/1998
17	PUSA-605 (MH-564)	425/E	08/06/1998
18	NANDI-8 (MH-741)	425/E	08/06/1998
19	JBV-2 (GKKV-93191)	425/E	08/06/1998
20	NANDI-32 (MH-773)	1050/E	26/10/1999
21	PUSA-415 (MP-739)	1050/E	26/10/1999
22	PUSA COMPOSITE-334 (MP-334)	1050/E	26/10/1999
23	MLBH-505 (MH-793, MLBH-44)	1050/E	26/10/1999
24	HHB-94	340/E	03/04/2000
25	HARIYANA COMPOSITE-10	340/E	03/04/2000
26	RHB-90 (MH-463)	821/E	13/09/2000

S.No	Crop/Variety	S.O.	Date
27	7688 (MH-795)	92/E	02/02/2001
28	PROAGRO-9443 (MH-846)	92/E	02/02/2001
29	GICKV-96752 (MP-363)	92/E	02/02/2001
30	NANDI-35 (MH-889)	1134/E	15/11/2001
31	PUSA COMPOSITE-383 (MP-383)	1134/E	15/11/2001
32	PROAGRO-9445 (MH-882-PB-112)	1134/E	15/11/2001
33	RHB-121 (MH-892)	1134/E	15/11/2001
34	COH (CO) 8	1134/E	15/11/2001
35	AIMP-92901 (SUMRUDHI-MP-282)	1134/E	15/11/2001
36	HC-20 (HMP 9102)	937/E	04/09/2002
37	GHB-558 (MH-946)	283/E	12/03/2003
38	MP-406 (CZP-9802)	283/E	12/03/2003
39	GHB-256 (MSH-105)	283/E	12/03/2003
40	HHB-146 (MH-960)	283/E	12/03/2003
41	GHB-577	161/E	04/02/2004
42	HHB-117	642/E	31/05/2004
43	PROAGRO-9444 (MSH-118)	122/E	02/02/2005
44	PARBHNI SAMPADA (PPC-6)	122/E	02/02/2005
	MAIZE / MAKKA (*ZEA MAYS*)		
1	KM-5981	360/E	01/05/1997
2	KM-5991	360/E	01/05/1997
3	JM-8	360/E	01/05/1997
4	PAC-9112 (PAC-91-P-12) (For 2 years only)	361/E	01/05/1997
5	PAC-9735 (PAC-35003) (For 2 years only)	361/E	01/05/1997
6	755 (For 2 years only)	361/E	01/05/1997
7	PAC-9703 (HA-003) (For 2 years only)	361/E	01/05/1997
8	3058 (Y-1402-K)	647/E	09/09/1997
9	JKMH-2492	647/E	09/09/1997
10	PRO-303 (3461)	647/E	09/09/1997
11	MMH-69	647/E	09/09/1997
12	HIM-129 (EHF-1121)	647/E	09/09/1997
13	KH-9451	647/E	09/09/1997
14	PRAKASH (JH-3189)	647/E	09/09/1997
15	DMH-1	647/E	09/09/1997
16	SSF-9374	662/E	17/09/1997
17	SHAKTI-1	662/E	17/09/1997
18	PUSA EARLY HYBRID MAKKA-1 (EH-202892)	662/E	17/09/1997

S.No	Crop/Variety	S.O.	Date
19	PUSA EARLY HYBRID MAKKA-2 (EH-203492)	662/E	17/09/1997
20	PRO-311 (4640)	662/E	17/09/1997
21	BIO-SEED-9681	662/E	17/09/1997
22	MMH-133	662/E	17/09/1997
23	COH-3	662/E	17/09/1997
24	3054 W (X-1382 W)	401/E	15/05/1998
25	3342 (X-1123-G)	401/E	15/05/1998
26	PAC-701 (ICI-701)	401/E	15/05/1998
27	PAC-705 (ICI-705)	401/E	15/05/1998
28	PAC- 9728 (For two years only)	402/E	15/05/1998
29	PAC –9747 (For two years only)	402/E	15/05/1998
30	32 A-09 (X-1174 W V)	425/E	08/06/1999
31	4642 (PRO-312)	425/E	08/06/1999
32	4210 (PRO-316) (For two years only)	425/E	08/06/1999
33	GAURAV (D-931)	425/E	08/06/1999
34	VIVEK HYBRID-4 (FH-3049)	425/E	08/06/1999
35	PAC-738 (AH-738)	425/E	08/06/1999
36	PAC-776 (AH-776)	425/E	08/06/1999
37	BIO-9673	425/E	08/06/1999
38	JAWAHAR COMPOSITE MAKKA-12 (JM-12)	425/E	08/06/1999
39	COBC-1 (BABY CORN)	425/E	08/06/1999
40	H.H.M.-1 (H.K.H.1082)	340/E	03/04/2000
41	H.H.M.-2 (H.K.H.1071)	340/E	03/04/2000
42	AMAR (D-941)	92/E	02/02/2001
43	JH-3459	92/E	02/02/2001
44	PUSA EARLY HYBRID MAKKA-3 (AH-58)	92/E	02/02/2001
45	VIVEK MAIZE HYBRID-9 (FH-3077)	92/E	02/02/2001
46	JKMH-175	92/E	02/02/2001
47	KOHINOOR	92/E	02/02/2001
48	VIVEK HYBRID-5	92/E	02/02/2001
49	NAC-6004	92/E	02/02/2001
50	ARAVALI MAKKA-1 (EV-90)	92/E	02/02/2001
51	GUJRAT MAKAI-4	92/E	02/02/2001
52	SEEDTEC-740 (SEEDTEC-2324)	1134/E	15/11/2001
53	SHAKTIMAN-1	1134/E	15/11/2001
54	NARMADA MOTI (IC-9001)	937/E	04/09/2002
55	PRIYA SWEETCORN	937/E	04/09/2002
56	DMH-2	937/E	04/09/2002

S.No	Crop/Variety	S.O.	Date
57	GUJRAT MAKAI-3	937/E	04/09/2002
58	NAC-6002	937/E	04/09/2002
59	JAWAHAR MAKKI-216 (JM-216)	937/E	04/09/2002
60	DK-701 (F-7001)	283/E	12/03/2003
61	GUJRAT MAKAI-6	283/E	12/03/2003
62	SHAKTIMAN-2	161/E	04/02/2004
63	SHEETAL	161/E	04/02/2004
64	JAWAHAR MAKKA-216	161/E	04/02/2004
65	PUSA EXTRA EARLY HYBRID MAKKA-5	161/E	04/02/2004
66	PRATAP HYBRID MAIZE-1 (EH-50802)	161/E	04/02/2004
67	D-994	642/E	31/05/2004
68	JKMH-222	642/E	31/05/2004
69	PRATAP MAKKA-3 (EC-3108)	122/E	02/02/2005
70	SUPER KOHINOOR (BISCO-2418)	122/E	02/02/2005
71	WIN ORANGE SWEET CORN	122/E	02/02/2005
72	BIO 9636 (BIO-92136)	122/E	02/02/2005
73	VL BABY CORN-1 (VL-78)	122/E	02/02/2005
74	DK-984 (2784)	122/E	02/02/2005
75	IC-8209 (72-A)	122/E	02/02/2005
76	VIVEK MAIZE HYBRID-15 (FH-3176)	122/E	02/02/2005
77	VIVEK MAIZE HYBRID-17 (FH-3186)	122/E	02/02/2005
78	AZAD KAMAL (R-9803)	122/E	02/02/2005
	PADDY / DHAN (*ORYZA SATIVA*)		
1	NIDHI (IET 9994)	360/E	01/05/1997
2	PUSA-677 (IET-12617)	360/E	01/05/1997
3	ASD-19 (IET-10436)	360/E	01/05/1997
4	ADT-42 (IET-13239)	360/E	01/05/1997
5	CORH-1	360/E	01/05/1997
6	PMK-2 (IET-13971)	360/E	01/05/1997
7	TPS-3	360/E	01/05/1997
8	SHYAMALA (IET-12561, R-259, WR-37-2)	360/E	01/05/1997
9	POORNIMA (IET-12284, R-281-PP-31-1)	360/E	01/05/1997
10	GR-6	360/E	01/05/1997
11	JITENDRA (IET-10526)	360/E	01/05/1997
12	PURNENDU (IET-10029)	360/E	01/05/1997
13	VIJETHA (MTU-1001)	647/E	09/09/1997
14	BASMATI-386	647/E	09/09/1997

S.No	Crop/Variety	S.O.	Date
15	COR-46 (TANU BPHR-8)	647/E	09/09/1997
16	ASD-20 (AS-89044, IET-11295)	647/E	09/09/1997
17	BARHAVRODHI (IET-11295)	647/E	09/09/1997
18	MALVIA DHAN-36	647/E	09/09/1997
19	PHB-71	647/E	09/09/1997
20	APHR-1	662/E	19/09/1997
21	APHR-2	662/E	19/09/1997
22	MDU-5	662/E	19/09/1997
23	ASWANI	662/E	19/09/1997
24	VLDHAN-61 (IET-13485)(VL-89-1179)	401/E	15/05/1998
25	TRIGUNA (IET-12875)	401/E	15/05/1998
26	JAWHARRICE-3-45 (IET-13623)	401/E	15/05/1998
27	IET-8116	401/E	15/05/1998
28	KRISHNA HAMSA	401/E	15/05/1998
29	DRRH-1	401/E	15/05/1998
30	KARNATAKA RICE HYBRID-2 (KRH-2)	401/E	15/05/1998
31	KESHAWA	401/E	15/05/1998
32	INDUR SAMBHA (PDR-763)	401/E	15/05/1998
33	SHIVA (WGL-3943)	401/E	15/05/1998
34	SONAMANI (CR-644) (IET-11365)	401/E	15/05/1998
35	TAPASWINI (IET-9945)	401/E	15/05/1998
36	DHOLA HEERA (IET-11411)	401/E	15/05/1998
37	RADHI (CRH-40) (IET-12413)	401/E	15/05/1998
38	NARENDRA USAR-2	401/E	15/05/1998
39	LUIT (TTB-127-216-2/IET-13622)	401/E	15/05/1998
40	JALA PRABHA	401/E	15/05/1998
41	SARASWATI	401/E	15/05/1998
42	KHANIKA	401/E	15/05/1998
43	JAMINI	401/E	15/05/1998
44	VLDHAN-81 (IET-13792)	425/E	08/06/1999
45	POOJA (IET-12241)	425/E	08/06/1999
46	CSR-13 (IET-10348)	425/E	08/06/1999
47	CORH-2	425/E	08/06/1999
48	ADTRH-1 (ADRTH-4)	425/E	08/06/1999
49	ADT-43 (IET-14878)	425/E	08/06/1999
50	TKM-11 (TM-6012)	425/E	08/06/1999
51	PANT SANKAR DHAN-1	425/E	08/06/1999
52	VAGAD DHAN	425/E	08/06/1999

S.No	Crop/Variety	S.O.	Date
53	WGL-3962	425/E	08/06/1999
54	GURJARI	425/E	08/06/1999
55	NARENDRA SANKAR DHAN-2	425/E	08/06/1999
56	LIEMA PHOU	425/E	08/06/1999
57	AKUT PHOU	425/E	08/06/1999
58	LALITA GIRI (IET-13198)	425/E	08/06/1999
59	UDAYA GIRI (IET-12136)	1033/E	13/10/1999
60	KHARVELA (IET-13253)	1033/E	13/10/1999
61	SEBATI (IET-11786)	1033/E	13/10/1999
62	BHOI (IET-12443)	1033/E	13/10/1999
63	GAJPATI (IET-13251)	1033/E	13/10/1999
64	KONARK (IET-12734)	1033/E	13/10/1999
65	SURENDRA (IET-12815)	1033/E	13/10/1999
66	MAHANADI (IET-13356)	1033/E	13/10/1999
67	INDRABATI (IET-13396)	1033/E	13/10/1999
68	PRACHI (IET-12786)	1033/E	13/10/1999
69	RAMA CHANDI (IET-13354)	1033/E	13/10/1999
70	CSR-27 (IET-13765, CSR-88, IR-6)	340/E	03/04/2000
71	HKR-46	821/E	13/09/2000
72	SOMASILA (NLR-33358)	821/E	13/09/2000
73	VEDAGIRI (NLR-33641)	821/E	13/09/2000
74	MARUTERU SANNALU (MTU-1006) IET-14348	821/E	13/09/2000
75	COTTONDORA SANNALU (MTU-1010)	821/E	13/09/2000
76	DEPTI (MTU-4870)	821/E	13/09/2000
77	VASUNDHARA (RGL-2538)	821/E	13/09/2000
78	CO-47 (IET-14298)	821/E	13/09/2000
79	SRIKAKULAM SANNAL (RGL-2537)	821/E	13/09/2000
80	GOLAK (IET-11898)	821/E	13/09/2000
81	SUNIL	821/E	13/09/2000
82	SUDHIR (IET-10543)	821/E	13/09/2000
83	NEERAJA (IET-11865)	821/E	13/09/2000
84	BHAGIRATHI (IET-11272)	821/E	13/09/2000
85	MAHANANDA	821/E	13/09/2000
86	SASHI	821/E	13/09/2000
87	SHATABDI (IET-4786)	821/E	13/09/2000
88	PNR-519	821/E	13/09/2000
89	SAHYADRI	821/E	13/09/2000
90	ADT-44	821/E	13/09/2000

S.No	Crop/Variety	S.O.	Date
91	PARAG-401 (PBNR-90-3-401)	821/E	13/09/2000
92	VIDEKDHAN-62 (IET-14621)	92/E	02/02/2001
93	6201 (CPA-103)	92/E	02/02/2001
94	NARENDER USAR-3	92/E	02/02/2001
95	KARNATAKA HILL PADDY-5 (KHP-5)	92/E	02/02/2001
96	HEMAVATI (DWR-4107)	92/E	02/02/2001
97	SRAVANI (NLR-33359)	92/E	02/02/2001
98	SWATHI (NLR-33057)	92/E	02/02/2001
99	BHARANI (NLR-30491)	92/E	02/02/2001
100	EARLY SAMBHA (RNRM-7)	92/E	02/02/2001
101	SURYA (BPT-4358)	92/E	02/02/2001
102	PY-6 (IET-11898)	92/E	02/02/2001
103	TRY-1	92/E	02/02/2001
104	PR-113	276/E	26/03/2001
105	PR-114	276/E	26/03/2001
106	PR-115	276/E	26/03/2001
107	PR-116	276/E	26/03/2001
108	PANT DHAN-16	1134/E	15/11/2001
109	VIVEKDHAN –82	1134/E	15/11/2001
110	HYBRID-6444	1134/E	15/11/2001
111	PUSA SUGANDHA-2	1134/E	15/11/2001
112	PUSA SUGANDHA-3	1134/E	15/11/2001
113	PUSA RH-10	1134/E	15/11/2001
114	VASUMATI	1134/E	15/11/2001
115	CSR-30 (YAMINI)	1134/E	15/11/2001
116	TRY-2	1134/E	15/11/2001
117	BAMLESHWARI	1134/E	15/11/2001
118	DANTASHWARI	1134/E	15/11/2001
119	CR 683-123	1134/E	15/11/2001
120	ADT (R)-45	1134/E	15/11/2001
121	ANJALI	937/E	04/09/2002
122	KAUM 57-18-1-1 (KRISHANJANA)	937/E	04/09/2002
123	RM-1 (MANAGLA MAHSURI)	937/E	04/09/2002
124	KAU-87117 (KARUNA)	937/E	04/09/2002
125	DEEPTI (WND-111)	937/E	04/09/2002
126	KAUM-57-9-1-1 (K-16)	937/E	04/09/2002
127	KAUM-45-20-1 (D-6)	937/E	04/09/2002
128	KAUM-42-6-3 (D-1) (MO-16-UMA)	937/E	04/09/2002

S.No	Crop/Variety	S.O.	Date
129	KAUM-20-19-4 (MO-15-RAMANIKA)	937/E	04/09/2002
130	KAUM-61-6-1-1-2 (GM-9)	937/E	04/09/2002
131	KAUM-59-29-2-1-2 (GM-1)	937/E	04/09/2002
132	MAKARAM (KTR-2)	937/E	04/09/2002
133	KAUA-4-4-2 (HARSHA)	937/E	04/09/2002
134	KUMBHAM (KTR-3)	937/E	04/09/2002
135	SARLA CR-260-77 (IET-10279)	937/E	04/09/2002
136	MUGAD SUGANDHA-1(IET-13549)	937/E	04/09/2002
137	VANDANA (RR-167-982)	937/E	04/09/2002
138	JAGABANDHU (IET-14100)	937/E	04/09/2002
139	DHANRASI (IET-15358)	283/E	12/03/2003
140	RH-204(EXPH-204)	283/E	12/03/2003
141	SHAH SARANG-1 (RCPL-1-87-8)	283/E	12/03/2003
142	LAM PNAH-1(RCPL-1-87-4)	283/E	12/03/2003
143	BHA LUM-1 (RCPL-1-27)	283/E	12/03/2003
144	BHALUM-2 (RCPL-1-29)	283/E	12/03/2003
145	GR-8	283/E	12/03/2003
146	DANDI (IET-14906)	283/E	12/03/2003
147	MTU-1031 (THOLAKARI)	161/E	04/02/2004
148	MTU-1032 (GODAVARI)	161/E	04/02/2004
149	SHANTHI (IET-12884)	161/E	04/02/2004
150	JGL-1798	161/E	04/02/2004
151	JGL-384	161/E	04/02/2004
152	CHRC (S) 2-1-7 (IET-13428)	161/E	04/02/2004
153	CSR-23 (IET-13769)	161/E	04/02/2004
154	HPR-1156 (IET-16007)	161/E	04/02/2004
155	PNR-546 (IET-11347)	161/E	04/02/2004
156	KALI KHASA	161/E	04/02/2004
157	SWATI (IET-12888)	161/E	04/02/2004
158	GR-104	161/E	04/02/2004
159	GR-7	642/E	31/05/2004
160	HKR-126	122/E	02/02/2005
161	SYE-2001	122/E	02/02/2005
162	PALAM DHAN-957 (IET-13795)	122/E	02/02/2005
163	SHARAVATHI (IR-57773)	122/E	02/02/2005
164	JR-503 (RICHA) (IET-16783)	122/E	02/02/2005
165	PUSA SUGANDH-5 (IET-17021)	122/E	02/02/2005
166	SURUCHI-5401 (MPH-5401)	122/E	02/02/2005

S.No	Crop/Variety	S.O.	Date
167	SUNGANDHAMATI (IET-16775)	122/E	02/02/2005
168	GR-9	122/E	02/02/2005
169	PKV MAKARAND	122/E	02/02/2005
170	PVK-SKL-3-11-25-30-36	122/E	02/02/2005
171	INDIRA DHAN-1 (IET-15376) (R 636-405)	122/E	02/02/2005
	SORGHUM/ JOWAR (*SORGHUM BICOLOR*)		
1	PUSA CHARI (HYBRID-106)	360/E	01/05/1997
2	BSR-1	360/E	01/05/1997
3	GJ-40 (DUAL-PURPOSE)	360/E	01/05/1997
4	PANT CHARI-4 (FORAGE SORGHUM)	360/E	01/05/1997
5	MLSH-296 (MLSH-14)	647/E	09/09/1997
6	CHS-16 (SPH-723)	647/E	09/09/1997
7	CSV-13	647/E	09/09/1997
8	DSV-5 (GRS-1)	647/E	09/09/1997
9	APK-1	662/E	17/09/1997
10	PAC-501 (ICI-501)	401/E	15/05/1998
11	DSV-4	401/E	15/05/1998
12	JKSH-22 (JKSH-161)	425/E	08/06/1999
13	CSH-17 (SPH-660)	425/E	08/06/1999
14	SAFED MOTI (FSH-92079)	425/E	08/06/1999
15	CO-27	425/E	08/06/1999
16	PALEM SORGHUM HYBRID-1	425/E	08/06/1999
17	GJ-41	425/E	08/06/1999
18	JAWAHAR JOWAR-1041	425/E	08/06/1999
19	CSH-18 (SPH-960)	1050/E	26/10/1999
20	PANT CHARI-5 (UPFS-32)	1050/E	26/10/1999
21	CSV-216 (PHULE YASHODA) (SPV-1359)	821/E	13/09/2000
22	CSH-19 R (SPH-1010-R)	821/E	13/09/2000
23	NANDYAL TELLA JONNA-3 (NTJ-3)	821/E	13/09/2000
24	MAHABEEJ-7 (SPH-981)	821/E	13/09/2000
25	RSLG-262 (MAULEE)	821/E	13/09/2000
26	PARBHANI SWETA (PVK-801) SPV-13	821/E	13/09/2000
27	SPH-840	92/E	02/02/2001
28	PAIYUR-2	92/E	02/02/2001
29	CO (S)-28	1134/E	15/11/2001
30	CO (FS)-29	1134/E	15/11/2001
31	ADILABAD SORGHUM HYBRID-1	1134/E	15/11/2001

S.No	Crop/Variety	S.O.	Date
32	DSH-4R	937/E	04/09/2002
33	SPH-837 (HYBRID)	937/E	04/09/2002
34	PRATAP JAWAR-1430 (SPV-1430)	161/E	04/02/2004
35	K-11 (FODER CUM GRAIN TYPE)	161/E	04/02/2004
36	PARBHANI MOTI (PVR-396/SPV-1 [illegible] 1)	122/E	02/02/2005
37	GUJRAT FODDER SORGHUM-5	122/E	02/02/2005
	BARLEY (*HORDEUM VULGARE*)		
1	RD-2503	360/E	01/05/1997
2	RD-2508	360/E	01/05/1997
3	REKHA BCU-73	360/E	01/05/1997
4	GEETANJALI (K-1149)	360/E	01/05/1997
5	MANJULA K-329	360/E	01/05/1997
6	MALTI (DL-88)	401/E	15/05/1998
7	RITAMBHARA (K-551)	401/E	15/05/1998
8	HARITMA (K-560)	401/E	15/05/1998
9	PRAGATI (K-508)	401/E	15/05/1998
10	HARIT (HBL-276)	425/E	08/06/1999
11	RD-2552	340/E	03/04/2000
12	NARENDER BARLEY-1 (NDB-209)	92/E	02/02/2001
13	NARENDER BARLEY-2 (NDB-940)	92/E	02/02/2001
14	NARMADA (K-603)	92/E	02/02/2001
15	PRITTI (K-409)	92/E	02/02/2001
16	DWR-28 (MALT BARLEY)	937/E	04/09/2002
17	NARENDRA BARLEY-3 (NDB-1020)	937/E	04/09/2002
18	BH-393	937/E	04/09/2002
19	HIMADRI (BHS-352)	283/E	12/03/2003
20	RD-2592	161/E	04/02/2004
21	RD-2624	161/E	04/02/2004
22	NDB-1173	122/E	02/02/2005
23	VL BARLEY-56	122/E	02/02/2005
	WHEAT (*TRITICUM* SPP.)		
1	GANGA (HD-2643)	360/E	01/05/1997
2	AMAR (HW-2004)	360/E	01/05/1997
3	VIJAY (DDK-1001)	360/E	01/05/1997
4	PDW-233 (DURUM)	360/E	01/05/1997
5	MACS-2694 (DURUM)	360/E	01/05/1997

S.No	Crop/Variety	S.O.	Date
6	NIWA-34	360/E	01/05/1997
7	VIDISHA (DL-788-2)	647/E	09/09/1997
8	HP-1761	647/E	09/09/1997
9	RAJESHWARI (HP-1744)	647/E	09/09/1997
10	V.L. GEHUN-738 (VL-738)	647/E	09/09/1997
11	PBW-373	647/E	09/09/1997
12	SWAPNIL (JWS-17)	662/E	17/09/1997
13	MACS-2846	401/E	15/05/1998
14	GW-273	401/E	15/05/1998
15	HS-365	401/E	15/05/1998
16	NARENDRA WHEAT-1012	401/E	15/05/1998
17	NARENDRA WHEAT-1014	401/E	15/05/1998
18	GANGA (DDK-1009)	401/E	15/05/1998
19	BHAVANI (HW-1085)	401/E	15/05/1998
20	GOMATI K-9465	401/E	15/05/1998
21	DWR-185	401/E	15/05/1998
22	SONAK	401/E	15/05/1998
23	SURBHI (HPW-89)	401/E	15/05/1998
24	UJIYAR (K-9006)	401/E	15/05/1998
25	MALAV SHAKTI (HI-8498)	425/E	08/06/1999
26	MALVIYA WHEAT 468 (HUW-468)	425/E	08/06/1999
27	SHRESTH (HD-2687)	425/E	08/06/1999
28	UP-2425	425/E	08/06/1999
29	UP-2382	425/E	08/06/1999
30	K.R.L.-19	340/E	03/04/2000
31	PBW-396	340/E	03/04/2000
32	PBW-443	340/E	03/04/2000
33	MALAV RATAN (HD-4672)	340/E	03/04/2000
34	K-9644	340/E	03/04/2000
35	H.I.-1454	340/E	03/04/2000
36	H.I.-1418	340/E	03/04/2000
37	H.W.-2044	340/E	03/04/2000
38	TEESTA (BW-1008)	821/E	13/09/2000
39	HD-2733 (VSM)	92/E	02/02/2001
40	MALVIYA WHEAT-510 (HUW-510)	92/E	02/02/2001
41	GANGOTRI-9162 (K-9162)	92/E	02/02/2001
42	HALNA (K-7903)	92/E	02/02/2001
43	PRASAD (K-8434)	92/E	02/02/2001

S.No	Crop/Variety	S.O.	Date
44	GW-1139	92/E	02/02/2001
45	KAUSHAMBI (HW-2045)	937/E	04/09/2002
46	ADITYA (HD-2781)	937/E	04/09/2002
47	GW-322	937/E	04/09/2002
48	VL-804	937/E	04/09/2002
49	NIAW-301 (TRIMBAK)	937/E	04/09/2002
50	WH-912	937/E	04/09/2002
51	WH-711	937/E	04/09/2002
52	SHIVALIK (HS-420)	283/E	12/03/2003
53	HIMGIRI (HS-375)	283/E	12/03/2003
54	AMRITA (HI-1500)	283/E	12/03/2003
55	DBW-14	283/E	12/03/2003
56	NARENDRA WHEAT-2036	283/E	12/03/2003
57	MP-4010	283/E	12/03/2003
58	VL GEHUN-829	283/E	12/03/2003
59	PDW-274 (DURUM)	283/E	12/03/2003
60	MACS-3125	283/E	12/03/2003
61	SUDHA (MPO(JW)-1106)	283/E	12/03/2003
107	RAJ-3777	283/E	12/03/2003
108	SWARNA (HI-1479)	283/E	12/03/2003
109	PBW-502	161/E	04/02/2004
110	POORVA (HD-2824)	161/E	04/02/2004
111	RAJ-4037	161/E	04/02/2004
112	VL GEHUN-832	161/E	04/02/2004
113	PALAM (HPW-147)	122/E	02/02/2005
114	CHANDRIKA (HPW-184)	122/E	02/02/2005
115	VL GEHUN-802	122/E	02/02/2005
116	URJA (HD-2864)	122/E	02/02/2005
117	MACS-6145	122/E	02/02/2005
	BLACK GRAM / URD (*VIGNA MUNGO*)		
1	KBG-512	360/E	01/05/1997
2	VAMBAN-2	662/E	17/09/1997
3	SHEKHAR-1 (KU-301)	401/E	15/05/1998
4	UG-218	401/E	15/05/1998
5	LAM-648 (LBG-648)	401/E	15/05/1998
6	TROMBAY UDID-94-2 (TU-94-2)	425/E	08/06/1999
7	WARANGAL-26	425/E	08/06/1999

S.No	Crop/Variety	S.O.	Date
8	BARK HA (RBU-38)	1050/E	26/10/1999
9	AZAD URD-1 (KU-91-2)	1050/E	26/10/1999
10	I.P.U. 94-1 (UTTRA)	340/E	03/04/2000
11	VAMBAN-3	821/E	13/09/2000
12	APK-1	92/E	02/02/2001
13	LAM-623 (LBG-623)	92/E	02/02/2001
14	LAM-685 (LBG-685)	92/E	02/02/2001
15	SHEKHAR-2 (KU-300)	1134/E	15/11/2001
16	KU-91 (AZAD URD-2)	1134/E	15/11/2001
17	SUMAN-JANA	937/E	04/09/2002
18	GOUTAM (WBU-105)	161/E	04/02/2004
19	SHEKHAR (KU-309)	161/E	04/02/2004
	CLUSTER BEAN (*CYAMOPSIS TETRAGOANALOBA*)		
1	HARAYANA GUAR -365	401/E	15/05/1998
2	RGC-986	425/E	08/06/1999
3	RGC-1003	425/E	08/06/1999
4	BUNDEL GUAR-3 (IGFRI-1079-1)	1050/E	26/10/1999
5	RGC-1002	1050/E	26/10/1999
6	KANCHAN BAHAR (M-83)	1135/E	15/11/2001
7	RGC-1017 (ARID LEGUME)	937/E	04/09/2002
8	HG-563	642/E	31/05/2004
9	SURYA (RGM-112)	122/E	02/02/2005
	COWPEA / LOBIA/ YARD LONG BEAN (*VIGNA UNGUICULATA*)		
1	KONKAN FODDAR COWPEA-1 (DFC-1)	360/E	01/05/1997
2	HARIYANA LOBIA-88 (FODDER TYPE)	360/E	01/05/1997
3	VAMBAN-1	647/E	09/09/1997
4	PUSA SAMPADA (V-585)	425/E	08/06/1999
5	UPC-9202	425/E	08/06/1999
6	GUJRAT COWPEA-3 (GC-3)	425/E	08/06/1999
7	GUJRAT COWPEA-4 (GC-4)	425/E	08/06/1999
8	KBC-2	92/E	02/02/2001
9	RC-101	92/E	02/02/2001
10	CO-6	92/E	02/02/2001
11	DURGA KRANTI (RCV-7)	1135/E	15/11/2001
12	KONKANWALI (DPL-VS-20)	1135/E	15/11/2001
13	SHUBHRA	937/E	04/09/2002

S.No	Crop/Variety	S.O.	Date
14	UPC-607 (FODDER COWPEA)	283/E	12/03/2003
	GREEN GRAM / MOONG (*VIGNA RADIATA*)		
1	TARM-1	360/E	01/05/1997
2	TARM-18	360/E	01/05/1997
3	PANT MOONG-4 (UPM-92)	662/E	17/09/1997
4	MALVIA JYOTI (HUM-1)	425/E	08/06/1999
5	RMG-268 (DRUG)	425/E	08/06/1999
6	K-1	425/E	08/06/1999
7	PRAGYA	425/E	08/06/1999
8	EKASILA (WGG-37)	425/E	08/06/1999
9	PRATAP (SG-1)	425/E	08/06/1999
10	MALVIYA JAGARTI (H.U.M.-2)	340/E	03/04/2000
11	CO-6 (COGG-902)	821/E	13/09/2000
12	PUSA-9531	92/E	02/02/2001
13	PUSA VISHAL	92/E	02/02/2001
14	LAM-460	92/E	02/02/2001
15	RMG-344 (DHANU)	92/E	02/02/2001
16	PKV-MOONG-8802	92/E	02/02/2001
17	GANGA-8 (GANGOTRI)	1134/E	15/11/2001
18	PDM-139	1134/E	15/11/2001
19	VRM (GG-1)	1134/E	15/11/2001
20	VBN (GG-2)	1197/E	14/11/2002
21	HUM-6 (MALVIYA JANPRIYA)	1134/E	15/11/2001
22	BPMR-145	937/E	04/09/2002
23	VAIBHAV (PHULE M-9339)	937/E	04/09/2002
24	MALVIYA JANCHETNA (HUM-12)	283/E	12/03/2003
25	SML-668	283/E	12/03/2003
26	RMG-492	283/E	12/03/2003
27	GM-4	283/E	12/03/2003
28	KAMADEVA (OUM-11-5)	161/E	04/02/2004
29	OBGG-52	161/E	04/02/2004
30	MUSKAN (MH-96-1)	161/E	04/02/2004
	HORSE GRAM / KULTHI (*MACROTYLOMA UNIFLORUM*)		
1	PALEM-1	401/E	15/05/1998
2	PALEM-2	401/E	15/05/1998
3	ARJIA KULTHI-21 (AK-21)	425/E	08/06/1999

S.No	Crop/Variety	S.O.	Date
4	PAIYUR-2	425/E	08/06/1999
5	PHG-9	92/E	02/02/2001
6	PRATAP KULTHI-1 (AK-42)	122/E	02/02/2005
	KESHARI DAL		
1	RATAN (BIO-212)	647/E	09/09/1997
2	PRATEEK-GRASSPEA (KHESARI)	92/E	02/02/2001
	MOTH BEAN / KIDNEY BEAN (*VIGNA ACONTIFOLIA*)		
1	FMM-96	425/E	08/06/1999
2	MARU VARDAN (RMO-225)	1050/E	26/10/1999
3	CAZRI MOTH-1 (CZM-79)	1050/E	26/10/1999
4	MARU BAHAR (RMO-435)	937/E	04/09/2002
5	RMO-423	161/E	04/02/2004
6	CAZRI MOTH-3 (CZM-99)	122/E	02/02/2005
	PIGEON PEA (HYBRID) RED GRAM (*CAJANUS CAJAN*)		
1	COPH-1	360/E	01/05/1997
2	NARENDRA ARHAR-1 (NDA-882)	647/E	09/09/1997
3	COPH-2	647/E	09/09/1997
4	AMAR (KA-32-1)	647/E	09/09/1997
5	SARITA	401/E	15/05/1998
6	PARAS (H-82-1)	401/E	15/05/1998
7	AZAD K-91-25-M	425/E	08/06/1999
8	DURGA (ICPL-84031)	425/E	08/06/1999
9	MALVIYA VIKALAP (M.A.-3)	1050/E	26/10/1999
10	APK-1 (ARG-102)	821/E	13/09/2000
11	VAMBAN-2 (VRG-4)	821/E	13/09/2000
12	LAXMI (ICPL-85063)	92/E	02/02/2001
13	TS-3	92/E	02/02/2001
14	BSMR-853 (VAISHALI)	937/E	04/09/2002
15	SELECTION-31	937/E	04/09/2002
16	MALVIYA VIKAS (MA-6)	283/E	12/03/2003
17	PUSA-992	161/E	04/02/2004
18	GT-101	161/E	04/02/2004
19	MALAVIYA CHAMATKAR (MAL-13)	122/E	02/02/2005
	RAJMESH (FRENCH BEAN) (*PHASEOLUS VULGARIS*)		
1	TRILOKI RAJMESH	401/E	15/05/1998

S.No	Crop/Variety	S.O.	Date
2	YERCAUD-1	1052/E	26/10/1999
3	OOTY-1	1135/E	15/11/2001
4	PHULE-SUREKHA (KOF B-4)	348/E	19/04/2001
5	AMBER (HPR-96-4)	937/E	04/09/2002
6	VARUN (ACPR-94040)	937/E	04/09/2002
7	KAILASH (SRC-74)	122/E	02/02/2005
	BENGAL GRAM (*CICER ARIETINUM*)		
1	SAMARAT (GNG-469)	360/E	01/05/1997
2	PUSA-391 (BG-391)	360/E	01/05/1997
3	VISHAL PHULE (G-87207)	360/E	01/05/1997
4	PGD-3(GFG-89-133-GPF-133)	647/E	09/09/1997
5	ALOK (KGD-1168)	647/E	09/09/1997
6	PANT G-186	647/E	09/09/1997
7	KARNALCHANA-1(CSG-8962)	401/E	15/05/1998
8	DCP-92-3	401/E	15/05/1998
9	WCG-1 (SADBHAWANA)	401/E	15/05/1998
10	PUSA KABULI-1003 (BG-1003)	425/E	08/06/1999
11	JG-11	425/E	08/06/1999
12	DHARWAD PRAGATEE (BGD-72)	425/E	08/06/1999
13	CO-3	425/E	08/06/1999
14	CO-4	425/E	08/06/1999
15	GG-1	425/E	08/06/1999
16	JG-322	425/E	08/06/1999
17	WCG-2 (SURYA)	425/E	08/06/1999
18	JGG-1	425/E	08/06/1999
19	PUSA CHAMATKAR (B.G.1053)	340/E	03/04/2000
20	HARYANA CHANA-3 (H-86-18)	340/E	03/04/2000
21	GUJRAT GRAM-4 (GCP-105)	821/E	13/09/2000
22	GULAK-1 (GULABI AKOLA-1)	92/E	02/02/2001
23	PKV-KABUL-2	92/E	02/02/2001
24	KRANTI (ICCC-47)	92/E	02/02/2001
25	WCG-10 (PANT-G-10)	92/E	02/02/2001
26	JAWAHAR GRAM-16 (SAKI-9516)	92/E	02/02/2001
27	VAIBHAV (RG-9218)	1133/E	15/11/2001
28	VIRAT (PHULE G-95418)	937/E	04/09/2002
29	HARYANA KABULI-1 (HK-89-131)	937/E	04/09/2002
30	JAWAHAR GRAM-130 (JG-130)	937/E	04/09/2002

S.No	Crop/Variety	S.O.	Date
31	ANUBHAV (RSG-888)	100/E	28/01/2003
32	VIHAR (PHULE G-95311)	283/E	12/03/2003
33	JAWAHAR GRAM KABUL-1 (JKG-92337)	283/E	12/03/2003
34	PDG-4	283/E	12/03/2003
35	GUJRAT GRAM-2	283/E	12/03/2003
36	GNG-1292	283/E	12/03/2003
37	CSJD-884 (AKASH)	283/E	12/03/2003
38	HARAYANA KABULI CHANA-2 (HK-94-134)	122/E	02/02/2005
	FIELD PEA / GARDEN PEAS (*PISUM SATIVUM*)		
1	UTTARA (MFP-8909)	360/E	01/05/1997
2	KPMR-144-1	647/E	09/09/1997
3	JAYANTI (HFP-8712)	401/E	15/05/1998
4	SOLON NIROG	843/E	21/09/1998
5	NAREBDRA SABJI MATAR-2 (NDV-P-2)	843/E	21/09/1998
6	SWATI (KFPD-24)	425/E	08/06/1999
7	MALVIYA MATAR-15 (H.U.D.P.15)	340/E	03/04/2000
8	PUSA PRABHAT (DDR-23)	92/E	02/02/2001
9	AMBIKA (IM-9102)	92/E	02/02/2001
10	KPMR-522 (JAY)	1134/E	15/11/2001
11	KPMR-400 (INDRA)	1134/E	15/11/2001
12	SHUBHRA (IM-9101)	1134/E	15/11/2001
13	PUSA PANNA (DDR-27)	1134/E	15/11/2001
14	VIVEK MATER-8	1135/E	15/11/2001
15	ADRASH (IPF-99-25)	161/E	04/02/2004
	LENTIL / MASUR (*LENS CULINARIS*)		
1	PUSA VAIBHAV (L-4147)	360/E	01/05/1997
2	GARIMA (LH-82-6)	360/E	01/05/1997
3	NARENDRA MASOOR-3 (NDL-92-1)	647/E	09/09/1997
4	DPL-62 (SHERI)	662/E	17/09/1997
5	SUBRATA (WBL-58)	401/E	15/05/1998
6	JAWAHAR MASOOR-3 (SLC-6)	340/E	03/04/2000
7	VL-MASOOR-103	821/E	13/09/2000
8	NOORI (IPL-81)	92/E	02/02/2001
9	PANT LENTIL-5	92/E	02/02/2001
10	LL-699	283/E	12/03/2003

S.No	Crop/Variety	S.O.	Date
	FABA BEAN /HORSE BEAN		
1	VIKARANT (VH-82-1)	425/E	08/06/1999
	INDIA BEAN / FIELD BEAN (LAB PURPUREUS)		
1	GUJRAT WAL-1	425/E	08/06/1999
	RICE BEAN / RED BEAN (*VIGNA UNBELLATA*)		
1	PRR-2 (PRR-8901)	401/E	15/05/1998
2	KONKAN RICE BEAN-1	821/E	13/09/2000
3	RBL-6	92/E	02/02/2001
4	BIDHAN-1 (BC-15/K-1)	92/E	02/02/2001
5	RBL-6	161/E	04/02/2004
6	BIDHAN RICE BEAN-2 (KRB-4)	122/E	02/02/2005
	BARNYARD MILLET (SAVA) (*ECHINOCHLOA COLONA*)		
1	V.L.MADIRA-172 (V.L.172)	340/E	03/04/2000
2	V.L.MADIRA-181	1134/E	15/11/2001
	DODO MILLET		
1	VAMBAN-1 (MNV-20)	360/E	01/05/1997
	KODO MILLET / KODARA / VARAGU (*PASPALUM SCROBICULATUM*)		
1	KK-1	425/E	08/06/1999
2	JAWAHAR KODO-155 (RKB-155)	821/E	13/09/2000
3	JAWAHAR KODO-48 (DPS-48)	1134/E	15/11/2001
	FINGER MILLET (RAGI) (*ELEUSINE CORACANA*)		
1	CHAMPAVATHI (VR-708)	401/E	15/05/1998
2	MR-1	401/E	15/05/1998
3	GPU-28	401/E	15/05/1998
4	MARUTHI	425/E	08/06/1999
5	GPU-26	821/E	13/09/2000
6	SUVRA (OUAT-2)	1033/E	13/10/1999
7	BHAIRVI (BM-9-1)	1050/E	26/10/1999
8	CHILIKA(OEB-10)	1134/E	15/11/2001
9	GPU-45	1134/E	15/11/2001
10	L-5	937/E	04/09/2002
	LITTLE MILLET/SAMAI/KUTKI (*PANICUM SUMSTRENSE*		
1	TNAU-63	360/E	01/05/1997

S.No	Crop/Variety	S.O.	Date
2	CO-3	662/E	17/09/1997
3	PAIYUR-2	821/E	13/09/2000
4	TARINI (OLM-203)	92/E	02/02/2001
5	KOLAB (OLM-36)	1134/E	15/11/2001
6	SABARA(OLM-20)	161/E	04/02/2004
	FOXTAIL MILLET ITALIAN MILLET / KANGHI (*SETARIA ITALICA*)		
1	TNAU-43	360/E	01/05/1997
2	PANT SETERIA-4	425/E	08/06/1999
3	MEERA (SR-16)	92/E	02/02/2001
4	PRATAP KANGHI-1 (SR-51)	161/E	04/02/2004
	PROSO MILLET / COMMON MILLET (*PENICUM MILIACEAUM*)		
1	GRUP-8 (DHPM-1)	92/E	02/02/2001
2	GRUP-21	161/E	04/02/2004
	CASTOR / ARANDI (*RICINUS COMMUNIS*)		
1	GCH-5 (SHD-145)	647/E	09/09/1997
2	TMV-6	647/E	09/09/1997
3	DEEPTI (DCH-32)	401/E	15/05/1998
4	TMVCH-1	425/E	08/06/1999
5	KRANTI (PCS-4)	425/E	08/06/1999
6	GCH-6 (JHB-665)	1050/E	26/10/1999
7	DEEPAK (DCH-177) HYBRID	821/E	13/09/2000
8	RHC-1	937/E	04/09/2002
9	KIRAN (PCS-136)	161/E	04/02/2004
10	HARITHA (PCS-124)	161/E	04/02/2004
	G.NUT / PEA NUT (*ARACHIS HYPOGEA*)		
1	ALR-2 (ALG-56) BUNCH TYPE	360/E	01/05/1997
2	BSRG-1 (ICGV-86143)	360/E	01/05/1997
3	DSG-1	647/E	09/09/1997
4	R-8808 (APOORVA)	662/E	17/09/1997
5	VRI-4	662/E	17/09/1997
6	KADIRI-4	662/E	17/09/1997
7	JAGTIALI-88 (ICG-88)	401/E	15/05/1998
8	R-9251	401/E	15/05/1998
9	HNG-10	425/E	08/06/1999

S.No	Crop/Variety	S.O.	Date
10	PRAKASH (CSMG-884)	425/E	08/06/1999
11	GUJRAT GROUND NUT (GG-5)	425/E	08/06/1999
12	JAWAHAR G. NUT (JGN-3)	425/E	08/06/1999
13	ALR-3 (ALG-63)	425/E	08/06/1999
14	PHULE VYAS (IL-220)	821/E	13/09/2000
15	CO-3 (TNAU-256)	821/E	13/09/2000
16	MANJRA (LGN-2)	92/E	02/02/2001
17	TIRUPATI-4 (TCGS-30)	92/E	02/02/2001
18	GG-7 (GUJRAT GROUND NUT-7) (J-38)	1134/E	15/11/2001
19	CO (GN) 4 (CULTURE TNAU-269)	1134/E	15/11/2001
20	VRI (GN)-5	1134/E	15/11/2001
21	SNIGDHA	937/E	04/09/2002
22	SNEHA	937/E	04/09/2002
23	AK-159	937/E	04/09/2002
24	GG-14(JSP-28)	283/E	12/03/2003
25	GG-6	283/E	12/03/2003
26	TG-37-A	161/E	04/02/2004
27	TPG-41	161/E	04/02/2004
28	GPBD-4 (VIKAS)	642/E	31/05/2004
29	KADIRI-5	122/E	02/02/2005
30	KADIRI-6	122/E	02/02/2005
31	PRUTHA (DH-86)	122/E	02/02/2005
32	RG-382	122/E	02/02/2005
	SESAME / TIL (*SESAMUM INDICUM*)		
1	VARISV-1	360/E	01/05/1997
2	D.S.-1	647/E	09/09/1997
3	JT-55 (TKG-55)	425/E	08/06/1999
4	SWETHA TIL	425/E	08/06/1999
5	RT-127	92/E	02/02/2001
6	AKT-64	92/E	02/02/2001
7	JTS-8	1134/E	15/11/2001
8	SHEKHAR (SH-446)	1134/E	15/11/2001
9	CHANDANA (JCS-94)	161/E	04/02/2004
10	PRACHI (ORM-17)	161/E	04/02/2004
	SOYABEAN (*GLYCINE MAX*)		
1	PANT SOYABEAN-1042 (P.K.-1042)	360/E	01/05/1997

S.No	Crop/Variety	S.O.	Date
2	PANT SOYABEAN-1024 (P.K.-1024)	360/E	01/05/1997
3	POOJA(MAUS-2)	360/E	01/05/1997
4	CO SOYA-2	360/E	01/05/1997
5	SNEHA (KB-79)	360/E	01/05/1997
6	AHILYA -1 (NRC-2)	360/E	01/05/1997
7	AHILYA -2 (NRC-12)	360/E	01/05/1997
8	AHILYA-3 (NRC-7)	360/E	01/05/1997
9	SL-295	647/E	09/09/1997
10	PK-1029	647/E	09/09/1997
11	MACS-450	425/E	08/06/1999
12	JS-90-41	425/E	08/06/1999
13	V.L.S.-47	340/E	03/04/2000
14	PRASAD (MAUS-32)	821/E	13/09/2000
15	PARBHANI SONA (MAUS-47)	821/E	13/09/2000
16	PANT SOYA-1092	821/E	13/09/2000
17	AHILYA-4 (NRC-37)	92/E	02/02/2001
18	INDIRA SOYA-9	92/E	02/02/2001
19	HARA SOYA	92/E	02/02/2001
20	L.S.B.-1	92/E	02/02/2001
21	JS-93-05 (JAWAHAR SOYABEAN 93-05)	937/E	04/09/2002
22	PRATIKAR (MAUS-61)	937/E	04/09/2002
23	PREATISHTHA (MAUS-61-2)	937/E	04/09/2002
24	SAMRUDHI (MAUS-71)	937/E	04/09/2002
25	MAUS-81	161/E	04/02/2004
26	PALAM SOYA (P-30-1-1)	122/E	02/02/2005
27	TAMS-38	122/E	02/02/2005
	SUNFLOWER / SURYA MUKHI (*HELIANTHUS ANNUS*)		
1	TANU SUF-10	360/E	01/05/1997
2	PAC-1091 (FOR TWO YEARS ONLY)	361/E	01/05/1997
3	ADARSH-45 (FOR TWO YEARS ONLY)	361/E	01/05/1997
4	PAC-47 (FOR TWO YEARS ONLY)	361/E	01/05/1997
5	SUNGENO-85 (HYBRID)	401/E	15/05/1998
6	PAC-36	401/E	15/05/1998
7	PAC-1091	401/E	15/05/1998
8	DSH-1	401/E	15/05/1998
9	LS-11	425/E	08/06/1999
10	TCSH-1 (HYBRID)	821/E	13/09/2000

S.No	Crop/Variety	S.O.	Date
11	MLSFH-47 (AH-11-34)	92/E	02/02/2001
12	KBSH-44	100/E	28/01/2003
13	SH-416	161/E	04/02/2004
14	PSFH-118	161/E	04/02/2004
15	PROSUN-09 (PRO-009)	122/E	02/02/2005
16	KBSH-41	122/E	02/02/2005
17	KBSH-42	122/E	02/02/2005
18	DRSF-108	122/E	02/02/2005
	LINSEED/ALSI (*LINUM USITATISSIMUM*)		
1	SHIKHA	401/E	15/05/1998
2	JAWAHAR LINSEED-5	425/E	08/06/1999
3	RASHMI (LCK-9216)	1050/E	26/10/1999
4	PADMINI (LMH-62)	1050/E	26/10/1999
5	MEERA (R.L.-933)	340/E	03/04/2000
6	PARVATI (LMH-16-5)	340/E	03/04/2000
7	SHEELA (LCK-9211)	92/E	02/02/2001
8	SHEKHAR (LCK-9313)	92/E	02/02/2001
9	NL-97	92/E	02/02/2001
10	RL-914	937/E	04/09/2002
11	JLS-27 (SUYOG) (SLS-27)	161/E	04/02/2004
12	BINWA (KL-210)	122/E	02/02/2005
	GOBHI SARSON		
1	NEELAM	92/E	02/02/2001
	INDIAN MUSTARD / RAYA (*BRASSICA JUNCEA*)		
1	GUJRATI MUSTARD-2	360/E	01/05/1997
2	LAXMI (RH-8812)	360/E	01/05/1997
3	PCR-7 (RAJAT)	647/E	09/09/1997
4	PBR-97	662/E	17/09/1997
5	AGRANI (SEJ-2)	401/E	15/05/1998
6	KIRAN (PBC-9211) KARAN SARSON	401/E	15/05/1998
7	CS-52 (DIRA-343)	401/E	15/05/1998
8	VSL-5 (JAGNATH)	425/E	08/06/1999
9	JAWAHAR MUSTARD-1	425/E	08/06/1999
10	URVASHI (RK-9501)	92/E	02/02/2001
11	RCC-4 (MUSTARD)	92/E	02/02/2001
12	ARAVALI (RN-393)	1134/E	15/11/2001
13	NARENDRA AGETI RAI-4	1134/E	15/11/2001

S.No	Crop/Variety	S.O.	Date
14	BASANTI (RK-8501)	1134/E	15/11/2001
15	AGRANI (SEJ-2)	283/E	12/03/2003
16	MAYA (RK-9902)	283/E	12/03/2003
17	GEETA (RB-9901) (RB-24)	283/E	12/03/2003
18	PUSA SWARNIM (IGC-01)	283/E	12/03/2003
19	SWARM JYOTI(RH-9801)	283/E	12/03/2003
20	VASUNDHARA (RH-9304)	283/E	12/03/2003
21	KANTI (RK-9807)	283/E	12/03/2003
22	RGN-13	283/E	12/03/2003
23	CS-54 (CS-614-4-1-4)	122/E	02/02/2005
	VEGETABLE MUSTARD		
1	PUSA SAAG-1 (MR-704)	642/E	31/05/2004
	NIGER / RAMTIL (*GUIZOTIA ABYSSINICA*)		
1	PAIYUR-1	647/E	09/09/1997
2	JNC-6	937/E	04/09/2002
3	NRS-96-1	122/E	02/02/2005
4	BIRSA NIGER-2 (BNS-8)	122/E	02/02/2005
	INDIAN RAPESEED AND MUSTARD YELLOW SARSON (*BRASSICA* SPP.)		
1	HYOLA-401 (FOR 2 YEARS ONLY)	361/E	01/05/1997
2	HYOLA-PAC-401	92/E	02/02/2001
3	NARENDRA SARSON-2 (NDYS-2)	647/E	09/09/1997
4	JHUMKA (YSBNC-1)	401/E	15/05/1998
5	RAGINI (MYSL-203)	92/E	02/02/2001
6	KBS-3	92/E	02/02/2001
	SAFFLOWER / KUSUM (*CARTHAMUS TINTORIUS*)		
1	A-2	647/E	09/09/1997
2	DSM-129	401/E	15/05/1998
3	MKM-11	401/E	15/05/1998
4	JSI-73	425/E	08/06/1999
5	NARI-6	92/E	02/02/2001
6	NARI-NH-1 (PH-6) (NON-SPINY HYBRID)	937/E	04/09/2002
7	PHULE KUSUMA (JLSF-414)	122/E	02/02/2005
	TORIA (*BRASSICA CAMPESTRIS* VAR *TORIA*)		
1	JMT-689	360/E	01/05/1997
2	ORT (M) 2-4 (PARBATI)	1134/E	15/11/2001

S.No	Crop/Variety	S.O.	Date
3	ORT (M) 6-2 (ANURADHA)	937/E	04/09/2002
	BAJRA NAPIER HYBRID GRASS (*PENNISETUM AMERICANUM*)		
1	CO-3	662/E	17/09/1997
2	KM-1 (ACK-2)	161/E	04/02/2004
	DINANATH GRASS (*PENNISETUM PEDICELLATUM*)		
1	COD-1	360/E	01/05/1997
	GUINEA GRASS (*PANICUM MAXIMUM*)		
1	PGG-518	451/E	08/06/1999
2	CO-2	82/E	13/09/2000
3	PGG-616	92/E	02/02/2001
4	BUNDEL GUINEA-1 (JHGG 96-5)	122/E	02/02/2005
	TALL FESCUS GRASS (*FESTUCA ARUNDINACEA*)		
1	HIM-1	401/E	15/05/1998
2	HIMA-4	122/E	02/02/2005
	SETARIA GRASS (PIGEON GRASS / NANDI GRASS) SETARIA ANCEPS		
1	NANDI	2/E	03/01/2005
2	PSS-1(GOLDENTIMOTHY)	380/E	15/05/1990
3	SETARIA-92	122/E	02/02/2005
	BERSEEM / EGYPTIAN CLOVER (*TRIFOLIUM ALEXANDRINUM*)		
1	JHB-46	647/E	09/09/1997
2	BUNDEL BERSEEM -3	92/E	02/02/2001
	OAT (*AVENA SATIVA*)		
1	BUNDEL JAI-851 (JMO-851)	401/E	15/05/1998
2	HARYANA JAVI-8	401/E	15/05/1998
3	BUNDEL JAI-992 (JHO 992)	122/E	02/02/2005
	DAINCHA (*SESBAINA ACULEATA*)		
1	PANT DAINCHA-1	122/E	02/02/2005
2	DH-2	122/E	02/02/2005
	COTTON (*GOSSYPIUM* SPP.)		
1	NBHB-11	360/E	01/05/1997

S.No	Crop/Variety	S.O.	Date
2	SURABHI (VRS-7)	360/E	01/05/1997
3	SVPR-2 (TSH-289)	360/E	01/05/1997
4	JKHY-2 (HYBRID)	360/E	01/05/1997
5	LDH-11 (HYBRID)	360/E	01/05/1997
6	JAWAHAR TAPTI	360/E	01/05/1997
7	H-11	360/E	01/05/1997
8	H-1098	360/E	01/05/1997
9	SURUTHI	647/E	09/09/1997
10	OMSHANKAR (CSHH-29)	647/E	09/09/1997
11	PUSA-8-6	647/E	09/09/1997
12	F-1378	647/E	09/09/1997
13	DHH-11	647/E	09/09/1997
14	KC-2	647/E	09/09/1997
15	KASHINATH (NFHB-109)	401/E	15/05/1998
16	ANKUR-09 (WHH-09)	401/E	15/05/1998
17	AJIT (LHH-144)	401/E	15/05/1998
18	AAH-1 (DESI COTTON HYBRID-1)	425/E	08/06/1999
19	NARSHIMHA (NANDYAL-1325)	425/E	08/06/1999
20	R.C.H.-2	340/E	03/04/2000
21	H.D.-123 (DESI COTTON)	340/E	03/04/2000
22	L-603	821/E	13/09/2000
23	L-604	821/E	13/09/2000
24	LAHH-4-(A-13)	821/E	13/09/2000
25	ARAVINDA (NDL-2708)	821/E	13/09/2000
26	MCU-12 (TCH-1025)	821/E	13/09/2000
27	SVPR-3	821/E	13/09/2000
28	PARBHANI-316 (GANGA)	821/E	13/09/2000
29	VICH-9	821/E	13/09/2000
30	SUMANGALA (CWROK-165)	92/E	02/02/2001
31	ICMF-20	92/E	02/02/2001
32	RG-18	92/E	02/02/2001
33	RS-810	92/E	02/02/2001
34	SAHANA (JK-276-8-2)	92/E	02/02/2001
35	RAMPBS-155	92/E	02/02/2001
36	VAGAD KALYAN (RB-423)	92/E	02/02/2001
37	G-COT-18	92/E	02/02/2001
38	PKVHY-4 (CAHH-8)	92/E	02/02/2001
39	AKA-7 (AKA-8307)	92/E	02/02/2001

S.No	Crop/Variety	S.O.	Date
40	PRATIMA (CNH-120MB)	1134/E	15/11/2001
41	NCS-145-BUNNY (NCHH-145)	1134/E	15/11/2001
42	G.COT-21	1134/E	15/11/2001
43	VICH-5	937/E	04/09/2002
44	HHH-223	937/E	04/09/2002
45	H-1117	937/E	04/09/2002
46	RS-2013	937/E	04/09/2002
47	JAWAHAR KAPAS-4 (JK-4)	937/E	04/09/2002
48	PHULE-492 (RHH-0492)	937/E	04/09/2002
49	PHULE-388 (RHB-388)	937/E	04/09/2002
50	SWADESHI-1 (ADCH-1)	283/E	12/03/2003
51	NCS-207 MALLIKA (NCHH-207)	283/E	12/03/2003
52	F-1861	283/E	12/03/2003
53	G.COT. 19 (G.AM-31)	283/E	12/03/2003
54	G.COT.MDH-11 (GSGDH-2)	283/E	12/03/2003
55	GUJRAT COTTON-23 (GBHV-179)	283/E	12/03/2003
56	VEENA	161/E	04/02/2004
57	PKV-DH-1 (AKDH-7)	161/E	04/02/2004
58	PKV HY-5 (CAHH-99)	161/E	04/02/2004
59	PA-255 (PARBHANI TURAB)	161/E	04/02/2004
60	NH-545	161/E	04/02/2004
61	RBDV-7 (PRATAP KAPI-1)	122/E	02/02/2005
62	CICR-2 (CISAA-2) (GMS BASED HYBRID)	122/E	02/02/2005
63	SHRESTH (CSHH-198)	122/E	02/02/2005
64	PH-348 (YAMUNA)	122/E	02/02/2005
65	PA-402	122/E	02/02/2005
66	PHULEJA-794	122/E	02/02/2005
67	MAHABEEJ-106	122/E	02/02/2005
68	MAHABEEJ DH-986	122/E	02/02/2005
	TOSSA JUTE & JUTE (*CORCHORUS* SPP.)		
1	JRO-66 (PBO-6)	401/E	15/05/1998
2	JRC-698 (SHRABANTI-WHITE-PBC-6)	1052/E	26/10/1999
3	JRC-8432 (SHAKTI TOSSA-CO-32)	1052/E	26/10/1999
4	BIDHAN PAT-3 (D-110)	92/E	02/02/2001
5	BIDHAN PAT-1 (D-18)	92/E	02/02/2001
6	BIDHAN PAT-2 (D-90)	92/E	02/02/2001
7	JRO-128 (SURYA) (CO-28)	937/E	04/09/2002

S.No	Crop/Variety	S.O.	Date
8	SUBALA (S-19)	122/E	02/02/2005
9	JRC-80 (MITALI WHITE)	122/E	02/02/2005
	SUNHEMP (PATUA) (*CROTALORIA JUNCEA*)		
1	LAKSHMI	13	19/12/1978
2	NARENDRA SANAI-1	122/E	02/02/2005
3	SAILASH (SH-4)	122/E	02/02/2005
	MESTA (ROSE LLE) *HIBISCUS CANABINUS*		
1	NC-583	598	08/10/1974
2	NIRMAL (MT-150)	122/E	02/02/2005
	CHILLI / CAPISICUM (HOT & SWEET PEPPER)- SIMLA MIRCH		
1	GUJRATCHILLI-2	1052/E	26/10/1999
2	PLR-1	1052/E	26/10/1999
3	JAWAHAR MIRCH-283	1052/E	26/10/1999
4	KASHMIR LONG-1	1135/E	15/11/2001
5	PMK-1	1135/E	15/11/2001
6	CO.3	1135/E	15/11/2001
7	NISHAT-1	1135/E	15/11/2001
8	UTAKALAVA(BC-14-2)	642/E	31/05/2004
9	RCH-1	642/E	31/05/2004
	BHINDI/OKRA (*ABELMOSCHUS ESCULENTUS*)		
1	CO.3 (HYBRID-8)	1052/E	26/10/1999
2	SHITLA UPHAR (DVR-1)	1135/E	15/11/2001
3	SHITLA JYOTI (DVR-1)	1135/E	15/11/2001
4	AZAD BHINDI-1	1135/E	15/11/2001
5	GUJRAT OKRA-2	1135/E	15/11/2001
	BITTER GOURD / KARELA (*MOMORIDICA CHARANTIA*)		
1	PRITI (MC-84)	843/E	21/09/1998
2	PHULE GREEN GOLD (RHRBG-3)	348/E	19/04/2001
3	PHULE PRIYANKA (RHRBG-1)	348/E	19/04/2001
	BOTTLEGOURD / LAUKI (*LAGENARIA SICERARIA*)		
1	PANT SHANKAR LAUKI-1	1052/E	26/10/1999
2	PUNJAB LONG (LC-2-1)	1052/E	26/10/1999

S.No	Crop/Variety	S.O.	Date
	ASH GOURD (PETHA) (*BENINCASA CERIFERA*)		
1	PUSA UJWAL (DAG-1)	642/E	31/05/2004
	SPONGE GOURD (LUFFA CYLINDRICA)		
1	PUSA SNEHA	642/E	31/05/2004
	BRINJAL / EGG PLANT (*SOLANUM MELONGENA*)		
1	UTKAL MADHURI (BB-44)	843/E	21/09/1998
2	UTKAL KESRI (BB-26)	843/E	21/09/1998
3	JC-1	843/E	21/09/1998
4	JC-2	843/E	21/09/1998
5	BEGUN SWETHA (SM-6-6)	843/E	21/09/1998
6	PUSA ANKUR (DBSR-91)	1052/E	26/10/1999
7	AZAD (HYBRID)	1052/E	26/10/1999
8	KKM-1 (KSM-107)	1052/E	26/10/1999
9	PLR-1	1135/E	15/11/2001
10	GUJRAT BRINJAL LONG-1	448/E	21/04/2003
11	SWARNA MANI (CHBR-1)	642/E	31/05/2004
12	BRINJAL HYBRID-2 (GBH-2)	642/E	31/05/2004
13	UTAKAL JYOTI (BB-13)	642/E	31/05/2004
14	SWARNA PRATIBHA (CH-309)	642/E	31/05/2004
	CABBAGE / PATAGOBHI (*BRASSICA OLERACEA*)		
1	KINNER RED	1052/E	26/10/1999
	CAULIFLOWER / PHUL GOBHI (*BRASSICA OLERACEA* VAR. *BOTRYTIS*)		
1	PUSA SNOWBALI K-25	161/E	04/02/2004
2	PUSA SHARAD (SEL-309-1-2)	642/E	31/05/2004
3	PUSA MEGHNA (DC/98-2)	642/E	31/05/2004
	CARROT (DAUCAS CAROTA)		
1	OOTY-1	1052/E	26/10/1999
	CASSAVA / TAPIOCA (*MANIHOT ESCULENTA*)		
1	SHRI HARSHA	843/E	21/09/1998
	GARLIC (*ALLIUM SATIVUM*)		
1	YAMUNA SAFED	1052/E	26/10/1999
2	V.L.GARLIC-1 (VLG-7)	642/E	31/05/2004

S.No	Crop/Variety	S.O.	Date
	MUSKMELON / KHARBUZA (*CUCUMIS MELO*)		
1	MHY-3	1052/E	26/10/1999
2	RM-50	642/E	31/05/2004
3	MHY-5	642/E	31/05/2004
	WATER MELON / TARBOOJ (*CITRELLUS LANATUS*)		
1	DURGAPURA LAL (RW-177-3)	642/E	31/05/2004
	SNAP MELON		
1	PUSA SHANDAR (DSM-1)	642/E	31/05/2004
	ONION (*ALLIUM CEPA*)		
1	PUNJAB WHITE	843/E	21/09/1998
2	PHULE SUVARNA	348/E	19/04/2001
3	RO-1 (RAJASTHAN ONION-1)	642/E	31/05/2004
	PALAK / SPINACH (*SPINACIA OLERACEA*)		
1	PUNJAB GREEN	843/E	21/09/1998
2	OOTY-1	1052/E	26/10/1999
	POTATO (*SOLANUM TUBEROSUM*)		
1	KUFARI PUKHRAJ	843/E	21/09/1998
2	KUFARI CHIPSONA-1	843/E	21/09/1998
3	KUFARI CHIPSONA-2	843/E	21/09/1998
4	KUFARI GIRIRAJ (SM/85-45)	1052/E	26/10/1999
5	KUFARI ANAND (MS/182-717)	1052/E	26/10/1999
6	KUFARI KANCHAN	1135/E	15/11/2001
	SWEET POTATO / SAKARKANDI (*IPOMOEA BATATAS*)		
1	SHRI RATNA	843/E	21/09/1998
2	SHRI BHADRA	843/E	21/09/1998
3	CROSS-4 (WHITE)	1052/E	26/10/1999
4	KONKAN ASHWINI	1135/E	15/11/2001
	TOMATO (*LYCOPERSICUM ESCULENTUM*)		
1	UTKAL KUMARI (BT-10)	843/E	21/09/1998
2	SOLAN VAJRA	843/E	21/09/1998
3	TH-802	843/E	21/09/1998
4	MUKTI (L-79-5)	843/E	21/09/1998
5	VASUNDHARA (HYBRID-28)	1135/E	15/11/2001

S.No	Crop/Variety	S.O.	Date
6	PARBHANI YASHASHRI (SEL-14)	1135/E	15/11/2001
7	SWAPRNA BAIBHAY (CHTH-1)	642/E	31/05/2004
8	SANKRANTHI (TLB-11)	642/E	31/05/2004
9	NANDI (TLB-130)	642/E	31/05/2004
10	VYBHAT (TLB-182)	642/E	31/05/2004
	GRAIN AMARANTHUS (CHOLAI) (*AMARANTHUS* SPP.)		
1	PRA-2 (PRA-9101)	92/E	02/02/2001
2	GUJRAT AMARANTHUS-2 (GA-2)	937/E	04/09/2002
3	KONKAN DURANGI	642/E	31/05/2004
	BUCK WHEAT (*FAGOPYRUM TATARICUM*)		
1	PRB-1 (PRB-9001)	843/E	21/09/1998
2	UDAY (KBB-3)	92/E	02/02/2001
	OPIUM POPPY (*PAPAVER SOMNIFERUM*)		
1	JAWAHAR OPIMIUM-539	1052/E	26/10/1999
	SENJI / INDIAN CLOVER (*MELLILOTUS PARVIFLORA*)		
1	WHITE SENJI	401/E	15/05/1998
	RED CLOVER		
1	PRC-3	122/E	02/02/2005
	SUGAR CANE (*SACCHARUM OFFICINARUM*)		
1	C.O.PANT-90223	821/E	13/09/2000
2	UTTARA (CO-87044)	821/E	13/09/2000
3	BHAVANI (CO-86249)	821/E	13/09/2000
4	KALYANI (CO-87025)	821/E	13/09/2000
5	PRABHA (CO-85004)	821/E	13/09/2000
6	DHANUSH (CO-91010)	821/E	13/09/2000
7	NAYANA (CO-86032)	821/E	13/09/2000
8	BHIMA (CO-8371)	821/E	13/09/2000
9	MOTI (CO-87268)	821/E	13/09/2000
10	SARAYU (CO-S-87263)	821/E	13/09/2000
11	RASEELI (CO.S-91230)	821/E	13/09/2000
12	COM-7714 (KRISHNA) / COM-88121 (IN PENISULAR ZONE)	821/E	13/09/2000
13	PRAMOD (B.O.-128)	92/E	02/02/2001
14	HARYANA-92 (COH-92201)	92/E	02/02/2001
15	GANDAK (CO-89029)	1134/E	15/11/2001

S.No	Crop/Variety	S.O.	Date
16	RASBHARI (COSE-95422)	1134/E	15/11/2001
17	RAJBHOG (COSE-92423)	1134/E	15/11/2001
18	CO.S-767	1134/E	15/11/2001
19	MADHUMATHI	937/E	04/09/2002
20	SHYAMA (CO-94008)	161/E	04/02/2004
21	MAHURI (CO-394164)	161/E	04/02/2004
22	COSE-95255 (RACHNA)	642/E	31/05/2004
23	COSE-96234 (RASHMI)	642/E	31/05/2004
24	COSE-96434 (JALPARI)	642/E	31/05/2004
25	SWETA (COS-94270)	122/E	02/02/2005
	CASHEW (*ANACARDIUM OCCIDENTALE*)		
1	VENGURLA-7	1135/E	15/11/2001
2	VRI-4 (M-15/4)	1135/E	15/11/2001
	FENUGREEK (METHI) (*TRIGONELLA FOENUMGRAECUM*)		
1	GUJRAT METHI-1 (JE-102)	348/E	19/04/2001
2	CO.2	1135/E	15/11/2001
	TARAN\MIRA / ROCKET SALAD (*ERUCASATIVA*)		
1	KARAN TARA (RTM-3140)	1134/E	15/11/2001
	CARDAMOM (*ELETTARIA CARDAMOMUM*)		
1	APPANGALA-1	1052/E	26/10/1999
	PUMPKIN / KASHIPHAL (*CUCURBITA MOSCHATA*)		
1	SOORAJ (CM-350)	448/E	21/04/2003
	CUCUMBER (KHEERA) (*CUCUMIS SATIVUS*)		
1	SWARNA POOMA (CH-20)	642/E	31/05/2004
	COLOCASSIA (TARO/ARVI) (*CALOCASIA ESCULANTA*)		
1	MUKTAKESHI	642/E	31/05/2004

DENOTIFICATION OF VARIETIES

S.No.	CROP/VARIETY	NOTIFICATION NO. & DATE	DENOTIFICATION NO. & DATE
	MAIZE		
1	GANGA-1	S.O.4045/24.9.69	S.O.661/E/17-9-97

S.No.	CROP/VARIETY	NOTIFICATION NO. & DATE	DENOTIFICATION NO. & DATE
2	GANGA-101	S.O.4045/24.9.69 & S.O.716/20.2.70	S.O.661/E/17-9-97
3	RANJEET MAKKA-1	S.O.4045/24.9.69	S.O.661/E/17-9-97
4	GANGA-3	S.O.4045/24.9.69 & S.O.716/20.2.70	S.O.661/E/17-9-97
5	MAIZE COMPOSITE JAWAHAR	S.O.4045/24.9.69	S.O.661/E/17-9-97
6	SONA	S.O.4045/24.9.69 & S.O.716/20.2.70	S.O.661/E/17-9-97
7	KISHAN	S.O.4045/24.9.69 & S.O.716/20.2.70	S.O.661/E/17-9-97
8	VIKRAM	S.O.4045/24.9.69 & S.O.5505/20.10.71	S.O.661/E/17-9-97
9	AMBER COMPOSITE	S.O.4045/24.9.69	S.O.661/E/17-9-97
10	GANGA-4	S.O.566(E)21.10.74	S.O.661/E/17-9-97
11	DECCAN-101 (EH-4207)	S.O.786/2.2.76	S.O.661/E/17-9-97
12	GANGA HYBRID MAKKA-9	S.O.470/19.2.80	S.O.661/E/17-9-97
13	HIM-128	S.O.470/19.2.80	S.O.661/E/17-9-97
14	K-1	S.O.19(E)14.1.82	S.O.424(E)/8.6.99
	PADDY		
1	AKASHI	S.O.1004/23.3.78	S.O.661/E/17-9-97
2	VANI	S.O.786/2.2.76	S.O.661/E/17-9-97
3	VIJAY	S.O.566(E)21.9.74	S.O.661/E/17-9-97
4	SONA	S.O.566(E)21.9.74	S.O.661/E/17-9-97
5	JAYANTI	S.O.566(E)21.9.74	S.O.661/E/17-9-97
6	KANCHI (CO-34)	S.O.786/2.2.76	S.O.661/E/17-9-97
7	BALA	S.O.566(E)21.9.74	S.O.661/E/17-9-97
8	PADMA	S.O.4045/24.9.69	S.O.661/E/17-9-97
9	CR-126-33-11(KALINGA-I)	S.O.440(E)/21.8.75	S.O.661/E/17-9-97
10	CR-126-42-2 (KALINGA-II)	S.O.440(E)/21.8.75	S.O.661/E/17-9-97
11	SUPHLA (CR-45-21-63)	S.O.295(E)/9.4.85	S.O.661/E/17-9-97
12	CR-36-148 (SUPRIYA)	S.O.440(E)/21.8.75	S.O.661/E/17-9-97
13	HEMA (CR-5-11)	S.O.440(E)/21.8.75	S.O.661/E/17-9-97
14	CO-36	S.O.13/19.12.78	S.O.424(E)/8.6.99
15	CO-44	S.O.596(E) 13.8.84	S.O.424(E)/8.6.99
16	CO-25	S.O.4045/24.9.69	S.O.424(E)/8.6.99
17	CO-29	S.O.4045/24.9.69	S.O.424(E)/8.6.99

S.No.	CROP/VARIETY	NOTIFICATION NO. & DATE	DENOTIFICATION NO. & DATE
18	CO-32	S.O.4045/24.9.69	S.O.424(E)/8.6.99
19	CO-41	S.O.19(E) 14.1.82	S.O.424(E)/8.6.99
20	CO-42	S.O.19(E) 14.1.82	S.O.424(E)/8.6.99
	BAJRA		
1	HB-1	S.O.4045/24.9.69 & S.O.716/20.2.70	S.O.661/E/17-9-97
2	HB-2	S.O.4045/24.9.69	S.O.661/E/17-9-97
3	HB-3	S.O.4045/24.9.69 & S.O.5505/20.10.71,	S.O.661/E/17-9-97
		S.O.440(E)/21.8.75 & 566(E)21.9.74	S.O.661/E/17-9-97
4	PHB-10	S.O.786/2.2.76	S.O.661/E/17-9-97
5	PHB14	S.O.786/2.2.76	S.O.661/E/17-9-97
6	BD-111	S.O.470/19.2.80	S.O.661/E/17-9-97
7	PUSA-763	S.O.371(E)29.5.82	S.O.661/E/17-9-97
8	MBH-110	S.O.2(E)3.1.83	S.O.661/E/17-9-97
9	MBH-130	S.O.165/6.3.87	S.O.661/E/17-9-97
10	CO-6	S.O.2(E)3.1.83	S.O.424(E)/8.6.99
	SORGHUM		
1	CSH-2	S.O.4045/24.9.69	S.O.661/E/17-9-97
2	CSH-3	S.O.566(E)21.9.74	S.O.661/E/17-9-97
3	CO-22	S.O.540(E)21.9.74	S.O.661/E/17-9-97
4	CO-23	S.O.19(E) 14.1.82	S.O.661/E/17-9-97
5	CO-24	S.O.19(E) 14.1.82	S.O.661/E/17-9-97
6	COH-3	S.O.832(E)19.11.85	S.O.661/E/17-9-97
7	CO-21	S.O.540(E)24.7.85	S.O.440(E)8.6.99
	COWPEA		
1	CO-3	S.O.19(E) 14.1.82	S.O.661/E/17-9-97
2	KM-1	S.O.19(E) 14.1.82	S.O.661/E/17-9-97
	REDGRAM		
1	CO-3	S.O.540(E)24.7.85	S.O.661/E/17-9-97
2	CO-4	S.O.596(E)13.8.84	S.O.424(E)/8.6.99
	GREEN GRAM		
1	CO-3	S.O.540(E)24.7.85	S.O.661/E/17-9-97

S.No.	CROP/VARIETY	NOTIFICATION NO. & DATE	DENOTIFICATION NO. & DATE
	BLACK GRAM		
1	CO-3	S.O.13/19.12.78	S.O.661/E/17-9-97
2	CO-4	S.O.2103/12.8.80	S.O.661/E/17-9-97
3	ADT-2	S.O.19(E) 14.1.82	S.O.661/E/17-9-97
	BENGAL GRAM		
1	CO-2	S.O.19(E) 14.1.82	S.O.424(E)/8.6.99
	CASTOR		
1	TMV-4	S.O.596(E)13.8.84	S.O.661/E/17-9-97
	GROUNDNUT		
1	CO-1	S.O.19(E) 14.1.82	S.O.424(E)/8.6.99
2	POL-2	S.O.13/19.12.78	S.O.424(E)/8.6.99
	COTTON		
1	TNB-1	S.O.596(E)13.8.84	S.O.424(E)/8.6.99
2	CSB-156	S.O.786/2.2.76	S.O.424(E)/8.6.99
3	MCU-8	S.O.786/2.2.76	S.O.424(E)/8.6.99
4	K-9 (KARUNGANNI)	S.O.19(E) 14.1.82	S.O.424(E)/8.6.99
5	G.COT 11	S.O.596(E)13.8.84	S.O.1133(E)/15.11.2001
6	G.COT-14	S.O.165(E) 6.3.87	S.O.1133(E)/15.11.2001
7	SANJAY (CJ-73)	S.O.19(E) 14.1.82	S.O.1133(E)/15.11.2001
8	NARBADA (A-51-9)	S.O.5505/20.10.71	S.O.123(E)/2.2.05
9	BADNAWAR-1	S.O.716/20.2.70	S.O.123(E)/2.2.05
10	MALJARI	S.O.716/20.2.70	S.O.123(E)/2.2.05
	SUN FLOWER		
1	CO-1	S.O.19(E) 14.1.82	S.O.424(E)/8.6.99

Reference: Notifiedvarieties and their Charactorstics RSSCA Pant Krishi Bhawan Publication.

Chapter - 20

Project Report on Establishment of A Seed Production/Processing Unit

The seed is a primary agricultural input. In fact every investment and efforts for getting better yields would go waste if the seed used is of poor quality. Moreover, a crop can survive without fertilizer, plant protection chemicals and even under extreme draught conditions but one can not think of a crop with out quality seed. The old saying "सुबीजम् सुक्षेत्रे जायते संवर्धते" goes true even today.

When we talk of a quality seed, we have to plan for a quality seed production, processing, handling, packing and storage. Keeping in mind a plant for production and processing of 5000 qtls. of seeds the project proposals are submitted.

Infrastructure required

To start with, the main crop Wheat can be taken in to consideration. The basic Infrastructure required for the purpose is :

Infrastructure (Basic Needs)

1. A land area of 200 hects. with an average yield potential of 25 qtls./ Hect. of Graded and Processed seed.
2. Identification of 40 contract seed growers willing to offer an area of 5 hects. per farmer on an average.
3. Foundation seed, 200 qtls. for 200 hects. of area specific improved high yielding varieties.
4. B t will bags 2000 for procurement of Raw seed.

Machinery and Equipments

Company Name:

Location:

Cost in lacs Rs.

1. Grader: 2.00

2. Gravity Separator:	0.50
3. Indent Cylinder:	0.50
4. Seed Treater:	0.20
5. Weighing Machine:	0.10
6. Bag closer:	0.08
7. Moisture tester	0.06
8. Elivator	0.50
9. Accessories	0.06
Total	**4.00**

Man power requirement

	No	Wages Rs./ Monthly	Total Investment Rs.
Executive Manager	1	8000.00	96000.00
Field Inspector & Processing Operator	1	6000.00	72000.00
Mechanic	1	3000.00	36000.00
Labour	6		50000.00
Watchman	2	2500.00	60000.00
		TOTAL	**314000.00**

Godown 3000 qtls. capacity

Capacity	3000	Cost of Land & Construction	10.00 Lac
		Total	**10.00 Lac**

Running Expenditure

Procurement Cost with Prem.	1011250.00
Cost of 60 qtls. F/S @ 1500/-	90,000.00
Handling Charges 1200 qtls.	3,000.00
Transportation	30,000.00
Packing Material G bags 2600 @ 20/-	52,000.00
Processing Material	10,000.00
Total	**1196250.00**

Cost Structure / Qtls. in Rs.

Rabi	Qty	Market rate	Premium	Rate	Total	Graded
Wheat						
Raj-3765	550	800	50	850	467500	440
C-306	250	1200	75	1275	318750	200
Raj-3077	500	700	50	750	375000	400
	1300	2700	175	2875	1161250	1040

	Raj-3765	C-306	Raj-3077
Procurement Cost	800.00	1200.00	700.00
Transportation & Handling	20.00	20.00	20.00
Prime Cost	820.00	1220.00	720.00
Processing Cost	25.00	25.00	25.00
Storage	12.00	12.00	12.00
Handling Charges	15.00	15.00	15.00
Publicity 1%	8.20	8.20	8.20
packing Charges	50.00	50.00	50.00
Stabilization 2.5%	20.50	20.50	20.50
X-Godown Cost	950.70 say 950	1350.70 say 1350	850.70 say 850
Profit 10%	95.00	135.00	85.00
Whole Sales Comm. 4%	38.00	53.00	33.00
Company price	**1083**	**1538**	**968**
Dealers Commissioner 8%	85.00	122.00	73.00
MRP	**1168**		
		1660	**1041**

Returns with 100% Sale of seed at Company Price : = 12,62,300.00

Returns from 250 Qt. undersize
@ 50% of price list i.e. 400/- & 550 = 1,11,500.00

Returns (Gross)

Returns	=	12,62,300.00
Returns with undersize	=	1,11,500.00
Returns with Sale of F/S	=	1,30,000.00
Gross Total returns	=	**1503800.00**

Total Expenditure

Management Cost	=	**314000.00**
Godown Rent 6 months	=	60,000.00
Depreciation on Equipment Cost	=	36,000.00
Running Expenses	=	**1196250.00**
Gross Profit : 15.03 – 11.96 lacs	=	307550.00
Repayment of loan with Interest	=	(–) 100000.00
NET PROFIT	=	**207550.00**

Additional Profit Source

Extra capacity utilization through custom grading 5000 qtls.

@ 20/- per qtl.	=	1,00,000.00
		307550.00

It is thus well understood that the proposals are economically viable and practically sound. There is hardly any fear of any failure. Since the farmer's seed need is unavoidable and presently when only 11% SRR is in existence there is enough scope in accordance with the GOI declaration to double the production by next decade. It is a fact that even if we could double the SRR, the contribution to total National Agricultural Produce would atleast get a thrust of 20%. Moreover, these proposals not only calm down our own thirst for employment, but puts us ourselves in a capacity to generate employment for atleast 10 more people.

The financial requirements for running expenses including procurement and handling would be met out through money circulation since the procurement of seeds would be spread over 12 months. Although the proposals have been prepared on "Wheat" basis but the crop equivalents can not be ignored. The major crops like Bajra, Maize, Jowar, Cotton and Castor etc. during Kharif season, Gram, Wheat, Mustard and Barley in Rabi in addition to vegetable seeds like Guar and others during summers leave immense scope for diversification of the activities and the plant could be run on 100% capacity utilization.

It is, therefore proposed that the Bank finances for processing equipments and for the constructions of processing plant and godown tune as follows

Machine cost	4.00 lac
Processing shed and land	10.00 lac
Raw seed purchase	10.00 lac
Total	23.50 lac

Some Important Addresses of Suppliers

(I) WEIGH-BRIDGE

1. M/S APPLE INSTRUMENTATION
   1-1, NEW MADHAVPURA MARKET,
   SHAHIBAUG ROAD,
   AHMEDABAD-380 004
   TEL- 91-79- 25633579, 25633580
   FAX - 91 - 79 - 25627559;
   CELL 91- 09824040199
2. LEOTRONIC SCALES PVT. LTD.,
   47, IST FLOOR, HIDE MARKET,
   AMRITSAR-143 001
   TEL- 0183 - 2550709, 2557844
   FAX- 91 - 183-2541104
3. JYOTI WEIGHING SYSTEMS LTD.
   1012-A, INDRA PRAKASH, 21,
   BARAKHAMBA ROAD,
   NEW DELHI-110 001
   PH. 011-41519020,0-9313232318
   FAX - 011- 23737450
4. M/S ENDEAVOUR INSTRUMENT P. LTD., 307, ANAND CHAMBERS,
   NR. OLD HIGH COURT RLY.
   CROSSING, NAVRANGPURA,
   AHMEDABAD- 380 009. GUJARAT.
   PH. 079-26440606/07
   FAX 91-079-26421221
5. PRECISION ELECTRONIC INSTRUMENTS CO.
   H-45, UDYOG NAGAR, PEERA GARHI, DELHI- 110041
   PH. 25962000, 25962017/18/19
   FAX- 91- 11-25479967

(II) D.G.SET

1. SHEKHAWATI ELECTRICAL ENGINEERS, B-22, SHAKTI BHAWAN, SHIVAJI COLONY
   KHATIPURA ROAD, JHOTWARA,
   JAIPUR
   TEL. 0141- 2341696 FAX- 2341551
2. DEEP RAIL NIRMAN PVT.LTD.
   110, CITY CENTRE, SANSAR CHANDRA ROAD,
   JAIPUR-302 001 (INDIA)
   TEL. 0141-2378897
3. NEW CHOUDHARY ENTERPRISES
   B.O. 306 CITY CENTRE, SANSAR CHANDRA ROAD,
   JAIPUR-302 001 (INDIA)
   PH. 3919139, M.- 9414037781

(III) MOISTURE METER

1. OSAW INDUSTRIAL PRODUCTS PVT. LTD., P.O. BOX NO. 42, OSAW COMPLEX, JAGADHRI ROAD
   AMBALA CANTT-133 001
   PH.-91 - 171-2699347,2699267
   FAX - 91 - 171- 2699102, 2699222

(IV) GRAVITY SEPERATOR 5 TPH

1. OSAW AGRO INDUSTRIES P. LTD.,
   P.O. BAG NO. 5, AGROSAW COMPLEX, JAGADHRI ROAD
   AMBALA CANTT-133 001
   PH.- 91 - 171-2699354, 2699167
   FAX - 91 - 171- 2699018

2. PADSONS INDUSTRIES PVT. LTD.
MANGALDAS MARKET,
AKOLA 444 001
PH. - 2258099, 6518634
FAX - 0724- 2259356

(V) 3 T.P.H. MACHINE LINE

1. OSAW AGRO INDUSTRIES P. LTD.
2. PADSONS INDUSTRIES P. LTD.
3. RADIANT EQUIPMENT COMPANY
14, RAJA PARK,
AMBALA CANTT- 133001,
HARYANA - INDIA
PH. 91-171-2891546, 2651547
M. 09315101546, 09416027547
FAX - 91 - 171- 2891545

(VI) PRE CLEANER CUM GRADER 5.00 T.P.H.

1. M/S RADIANT
2. OSAW AGRO
3. PADSONS

(VII) 5 TPH MACHINE LINE

1. OSAW AGRO
2. PADSONS

(VIII) PRE CLEANER CUM GRADER 7.00 T.P.H.

1. RADIANT
2. OSAW AGRO
3. PADSONS